Visualized Flow

Fluid motion in basic and engineering situations revealed by flow visualization

Karman vortex after two cylinders parallel to flow (p. 50).

Edited by
THE JAPAN SOCIETY OF MECHANICAL ENGINEERS

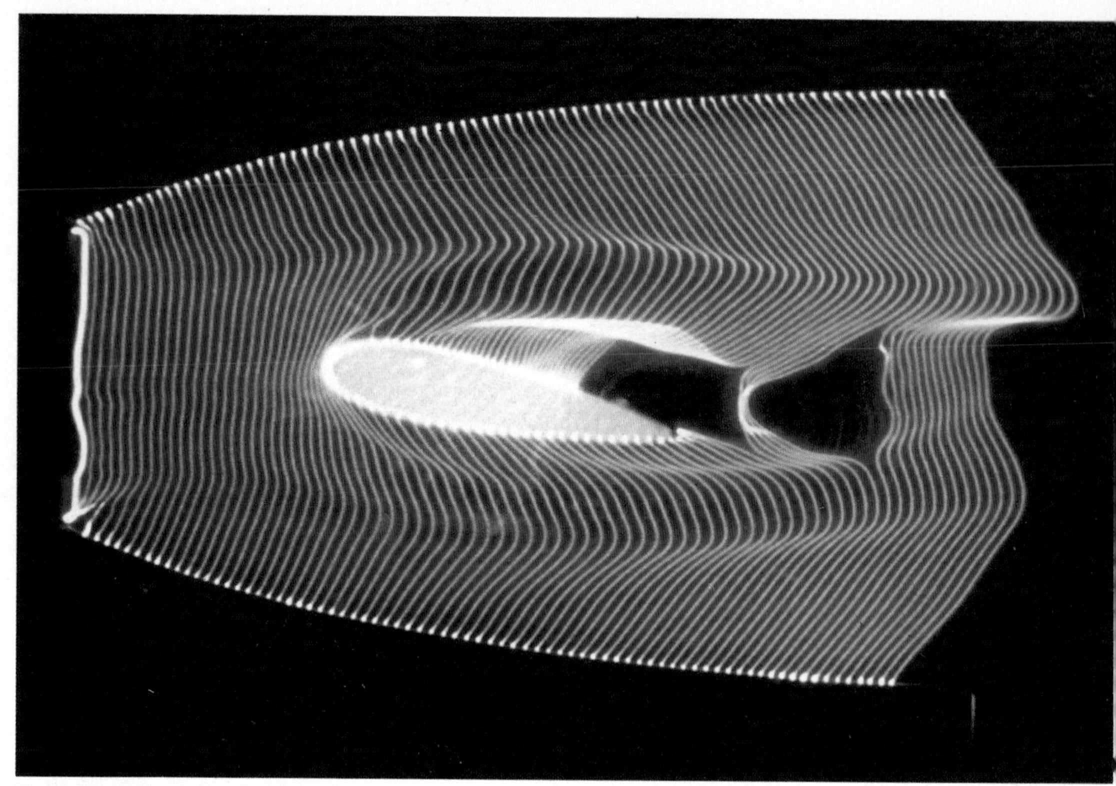

Flow around an aerofoil (air velocity 28 m/s, chord 40 mm, incidence 10° spark tracing method) (p. 87).

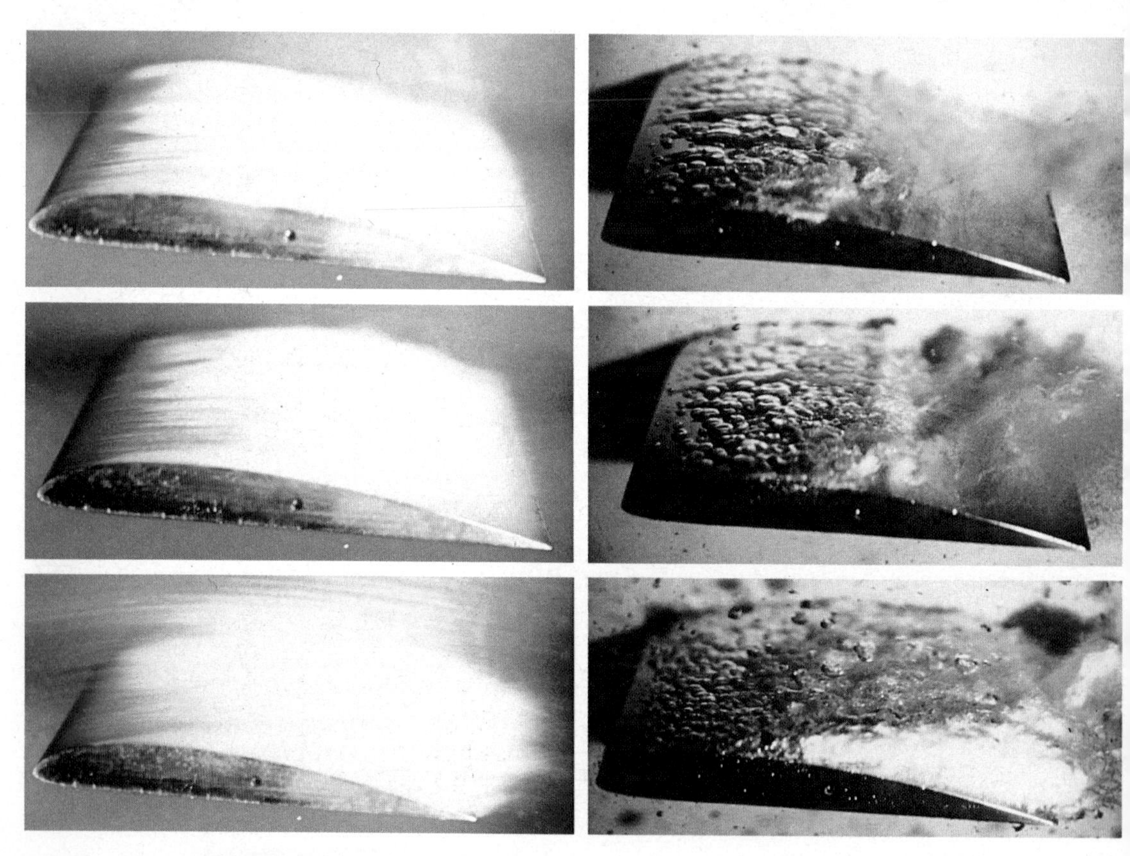

Cavitation generated on an aerofoil (p. 119).

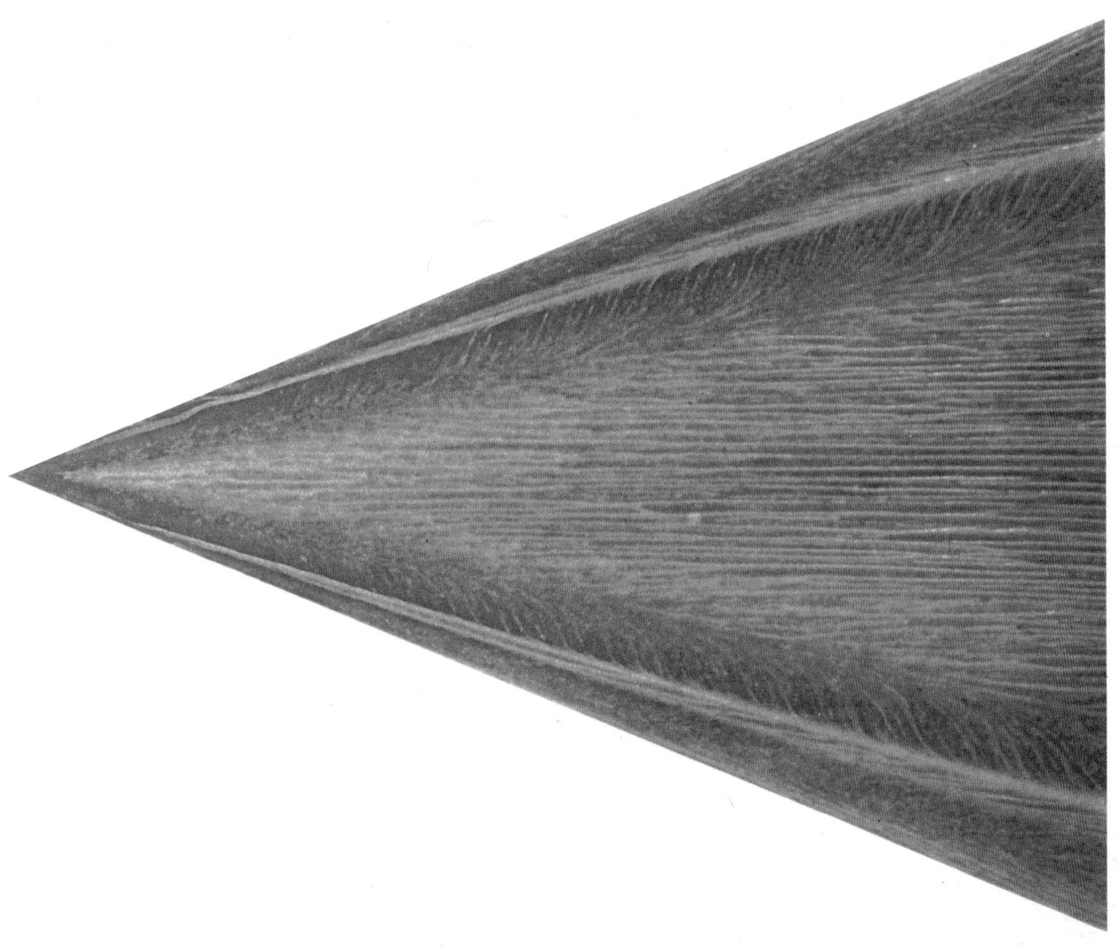
Oil flow pattern on a delta wing (p. 89).

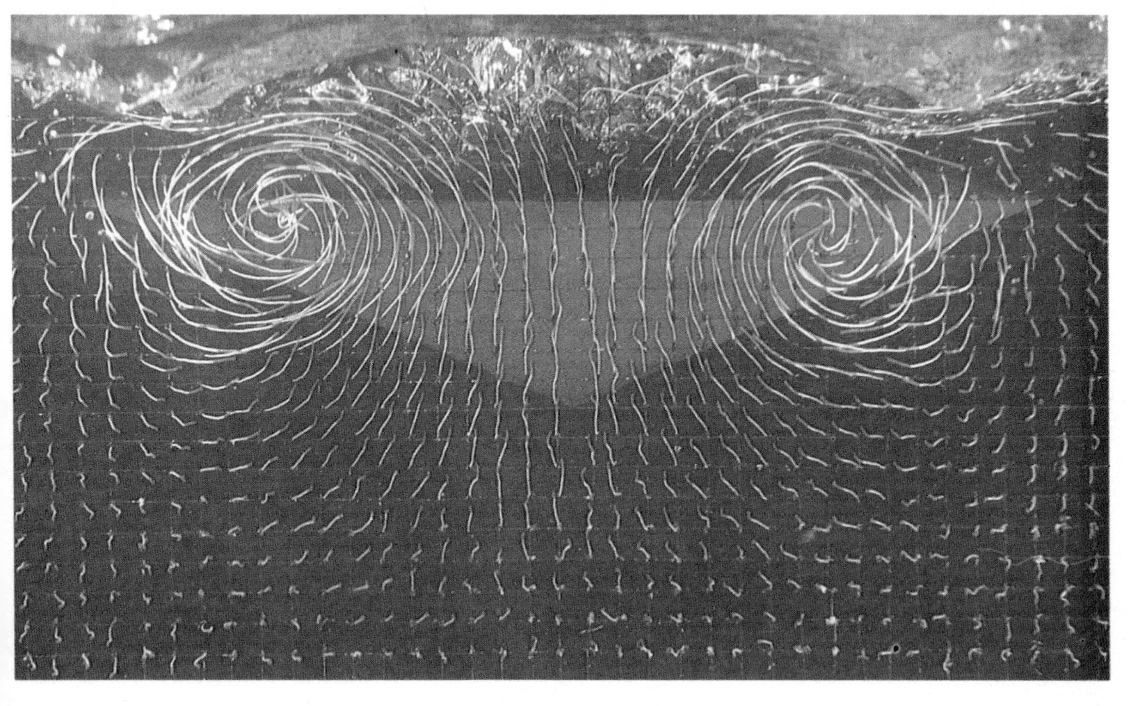
Wake of a delta wing (p. 59).

Flow around a streamlined car model (p. 63).

 Wake of an automobile (p. 64).

Flow in an impeller of a centrifugal pump (p. 97).

Compressible fluid flow around a wedge profile (p. 34).

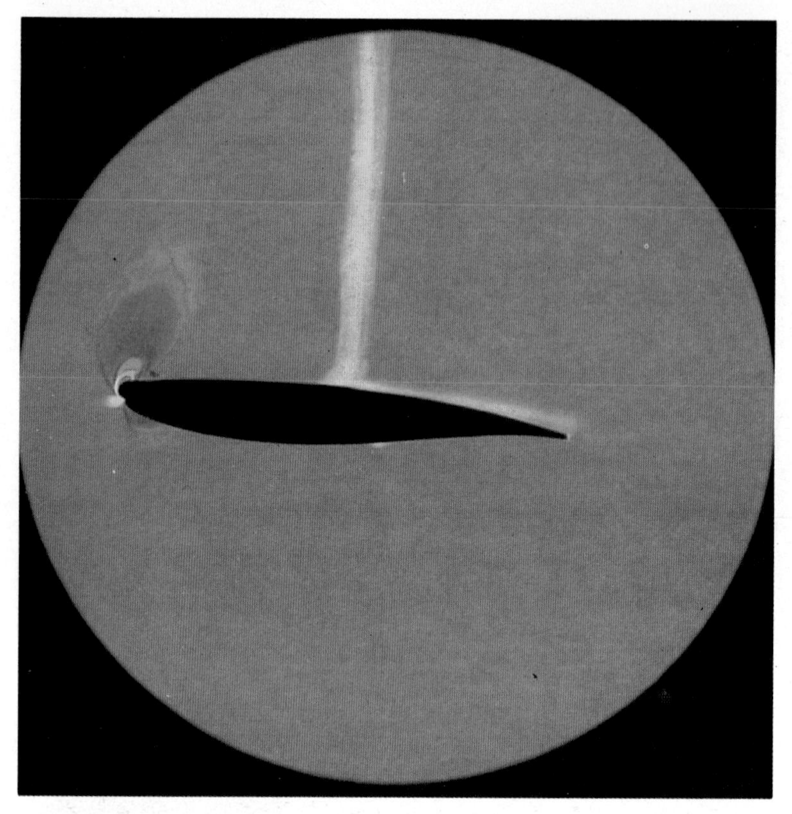

Transonic flow around a two-dimensional aerofoil (p. 86).

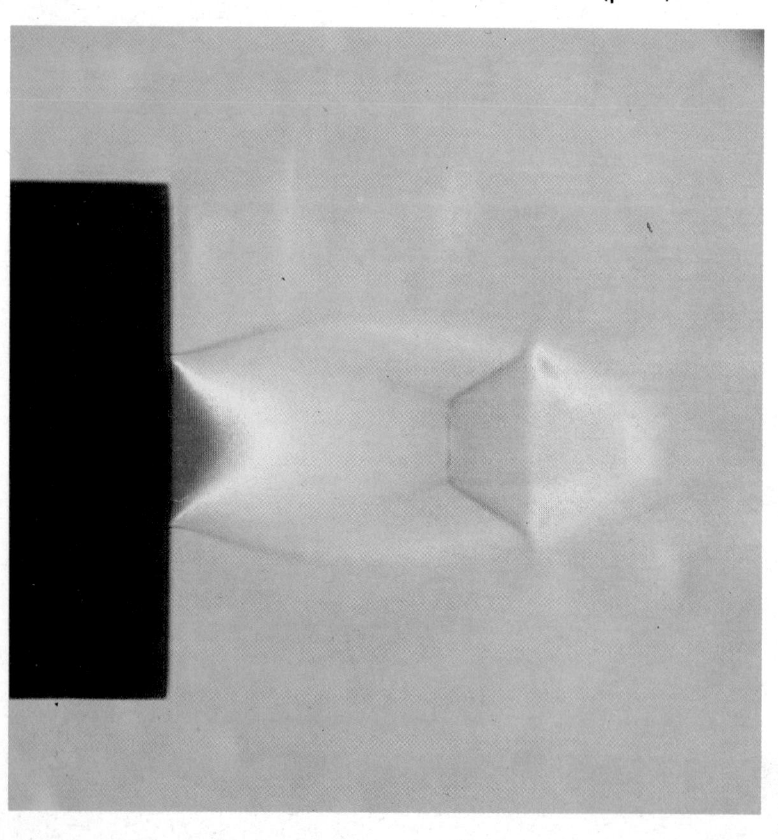

Supersonic jet from a nozzle (p. 47).

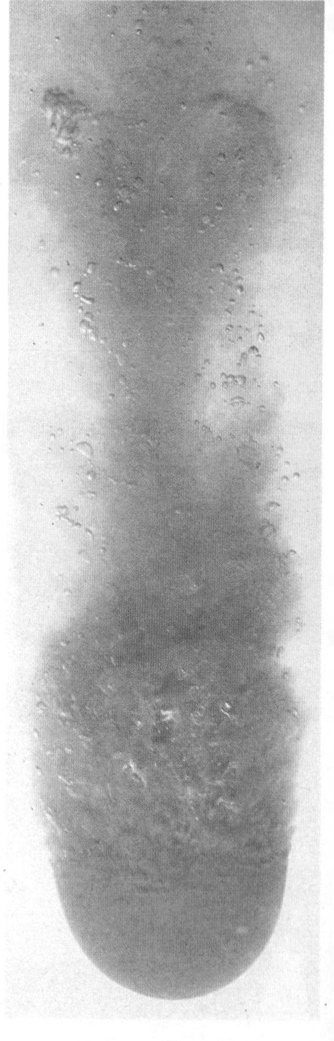

Wake of a dropping ball (p. 57).

Supersonic flow incoming to a jet engine intake (air, $M=2$, outer diameter of the intake 60 mm, $Re=10^7$, color-schlieren method) (p. 34).

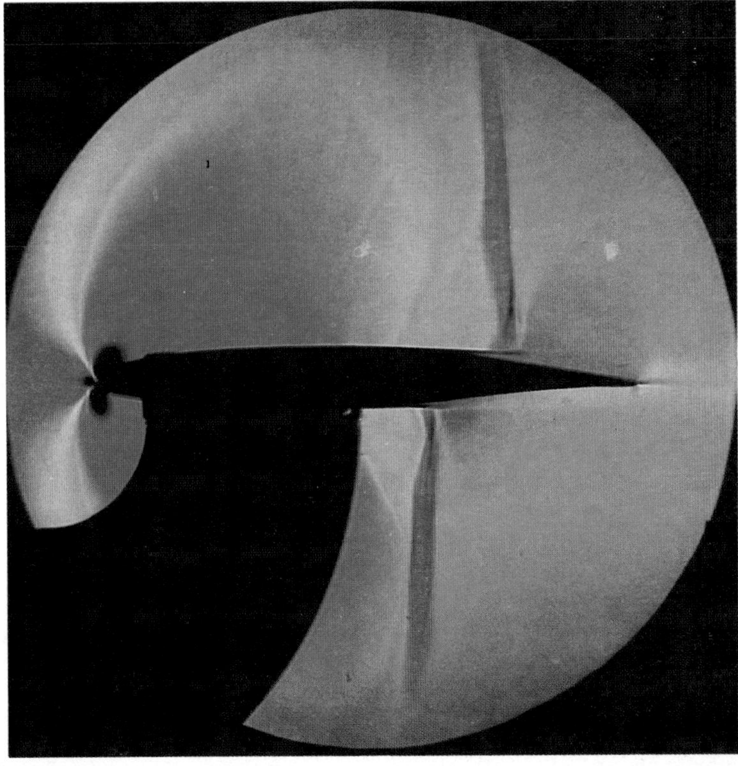

Transonic flow around an aerofoil (p. 39).

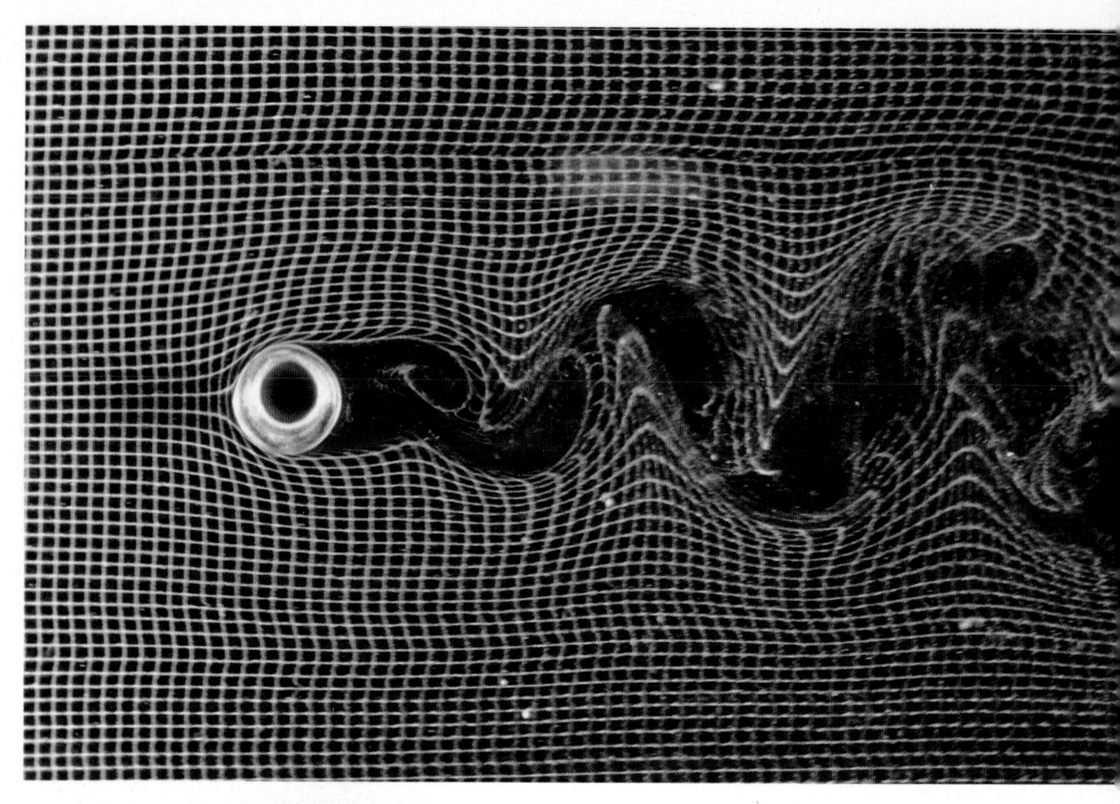

Karman vortex behind a circular cylinde
(p. 11).

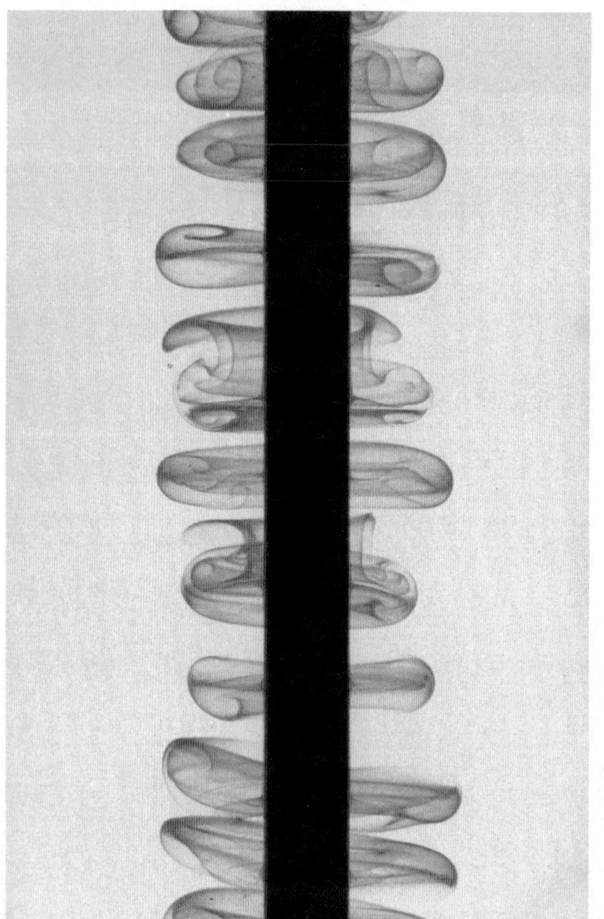

Vortex around a rotating circular cylinder
(p. 109).

Visualized Flow

Fluid motion in basic and engineering situations revealed by flow visualization

Compiled by
THE JAPAN SOCIETY OF MECHANICAL ENGINEERS

Chairman of Editorial Committee

Y. NAKAYAMA

Tokai University, Hiratsuka, Japan

UK Editors

W. A. WOODS

D. G. CLARK

Queen Mary College, University of London, UK

PERGAMON PRESS

OXFORD · NEW YORK · BEIJING · FRANKFURT
SÃO PAULO · SYDNEY · TOKYO · TORONTO

U.K.	Pergamon Press plc, Headington Hill Hall, Oxford OX3 0BW, England
U.S.A.	Pergamon Press, Inc., Maxwell House, Fairview Park, Elmsford, New York 10523, U.S.A.
PEOPLE'S REPUBLIC OF CHINA	Pergamon Press, Room 4037, Qianmen Hotel, Beijing, People's Republic of China
FEDERAL REPUBLIC OF GERMANY	Pergamon Press GmbH, Hammerweg 6, D-6242 Kronberg, Federal Republic of Germany
BRAZIL	Pergamon Editora Ltda, Rua Eça de Queiros, 346, CEP 04011, Paraiso, São Paulo, Brazil
AUSTRALIA	Pergamon Press Australia Pty Ltd., P.O. Box 544, Potts Point, N.S.W. 2011, Australia
JAPAN	Pergamon Press, 5th Floor, Matsuoka Central Building, 1-7-1 Nishishinjuku, Shinjuku-ku, Tokyo 160, Japan
CANADA	Pergamon Press Canada Ltd., Suite No. 271, 253 College Street, Toronto, Ontario, Canada M5T 1R5

Translated from the Japanese edition published by Maruzen in 1984

First English edition 1988

Library of Congress Cataloging in Publication Data

Visualized flow.
(Thermodynamics and fluid mechanics series)
1. Flow visualization. I. Nakayama, Y. (Yasuki)
II. Nihon Kikai Gakkai. III. Series.
TA357.V53 1987 620.1'064 87-21536

British Library Cataloguing in Publication Data

Visualized flow : fluid motion in basic and engineering situations. — (Thermodynamics and fluid mechanics series).
1. Fluid dynamics
I. Japan Society of Mechanical Engineers
II. Nakayama, Y. III. Woods, W.A.
IV. Clark, D.G. V. Series
532'.051 QC151

ISBN 0-08-034065-2 Hardcover
ISBN 0-08-034064-4 Flexicover

Printed in Great Britain by
Hazell Watson & Viney Limited
Member of BPCC plc
Aylesbury, Bucks, England

Foreword

THE BOOKS in the Thermodynamics and Fluid Mechanics division of the Pergamon International Library were originally planned as a series for undergraduates, to cover those subjects taught in a three-year course for Mechanical Engineers. Subsequently, the aims of the series were broadened and a number of volumes have been introduced which catered for undergraduates, post-graduate students and engineers in practice. These included new editions of books published earlier in the series.

The present volume continues this development and it will be useful to experienced engineers in industry and students new to the subject of fluid mechanics. The main purpose of the book is to provide information for engineers and students to improve their understanding of flow phenomena. The work presented here ranges from fundamental aspects, such as laminar and turbulent flow, to engineering applications; for example understanding why cavitation damage occurred on the runner of a Francis turbine.

The objectives and the range of engineering applications make this work different from, and perhaps complementary to, the excellent album of fluid motion assembled in 1982 by Milton Van Dyke. Inevitably there is a certain amount of overlap and this editor agrees with Milton Van Dyke that the subject of fluid mechanics cannot be arranged in a linear fashion. The layout followed here may be described, adopting a phrase, as a pleasant medley of fluid motion and applications.

This book presents the work of some fifty contributors drawn mainly from Academia and Industry in Japan as explained in the Preface and what follows. It was first published in Japanese in 1984, and although some changes have been made, the same general layout has been retained.

With more importance being given to Engineering Applications in engineering education as a result of the policies of the UK Engineering Council, the publication of this book in English is most timely.

W A WOODS
Series Editor

Preface

FLOW visualization has contributed greatly to the development of modern fluid dynamics and has been utilized by such pioneers as O. Reynolds (1842–1912) and L. Prandtl (1875–1953), for example. Various fluid flow phenomena (air or water) can also be observed in our daily life. Fluid mechanics is taught at college or university but the understanding of basic flow phenomena is not always easy. Flow visualization always plays an important role in understanding flow phenomena and provides physical-intuitive reasoning. This book was designed to enhance the understanding of basic flow phenomena by students or practicing engineers through flow visualization photographs and the corresponding explanations.

The book was originally edited by a committee of the Japan Society of Mechanical Engineers and published in Japanese in 1984. The English version has been edited again by the same committee in conjunction with the series editor and over 200 pictures are presented.

The contents of this album of flow phenomena revealed by various flow visualization techniques consist of: (1) basic flow phenomena, (2) laminar and turbulent flows, (3) similarity, (4) compressible flows, (5) jets, (6) external flow, (7) internal flow, (8) circulation theory, (9) flow around a wing, (10) internal flows in blade cascades and fluid machinery, (11) unsteady flow, (12) cavitation and (13) non-Newtonian fluid. For easy reference, an outline of flow visualization techniques is also given at the beginning.

It is hoped that this book will also be useful to engineers and scientists dealing with various flow problems in science, engineering, agriculture medical science, and physical education. The editorial committee wishes to thank Pergamon Press for agreeing to the publication of this book.

Yasuki Nakayama
Chairman of Editorial Committee

Contents

Publication Committee

Contributors

Hideaki AKAMATSU	Kyoto University
Hikaru ITO	Meiji University
Masao OSHIMA	Ebara Co., Ltd.
Eisuke OUTA	Waseda University
Nobuhide KASAGI	University of Tokyo
Hiroshi KATO	Tokyo Metropolitan University
Toshio KURIHARA	Tokyo Aerodynamic College
Junichi KUROKAWA	Yokohama National University
Hideo FURUSHIRO	Mitsubishi Heavy Industries, Ltd.
Jiro SAKAGAMI	Ochanomizu University
Norio SHINPO	Tokyo Aerodynamic College
Tetsuo TAGORI	University of Tokyo
Kiyoaki TAJIMA	Waseda University
Michitoshi TAKAGI	Nissan Co., Ltd.
Yoshimichi TANIDA	University of Tokyo
Sadatoshi TANEDA	Kyushu University
Yoshimasa TOMONARI	Nippon Bunri University
Tomitaro TOYOKURA	Yokohama National University
Hiroshi NAKAGUCHI	Emeritus Professor, University of Tokyo
Yasuki NAKAYAMA	Tokai University
Taku NAGATA	Gifu University
Takehiko NAGAYAMA	Mitsubishi Heavy Industries, Ltd.
Takenori NISHI	National Aerospace Laboratory
Shigeo FUJIKAWA	Kyoto University
Kazuyasu MATSUO	Kyushu University
Hitoshi MURAI	Hiroshimma Institute of Technology
Hideo YAMADA	Nagoya Institute of Technology
Masanobu YAMAMASU	Kanto Gakuin University
Toyoaki YOSHIDA	National Aerospace Laboratory

Presenters of Photographs

The head number indicates figure number

Frontispiece

iii	Toshio KOBAYASHI, University of Tokyo
iv-upper, x-upper	Yasuki NAKAYAMA, Tokai University
iv-lower, v-upper	Hitoshi MURAI, Hiroshima Institute of Technology
v-lower, vi-lower	Tetsuo TAGORI, University of Tokyo
vi-upper	Michitoshi TAKAGI, Nissan Co., Ltd.
vii-upper	Masanobu YAMAMASU, Kanto Gakuin University
vii-lower	Eisuke OTA and Takamasa OCHIAI, Waseda University
viii-upper	Takenori NISHI, Toshiyoshi FUJITA and Nobuhiko KAMIYA, National Aerospace Laboratory
viii-lower left	Nobutoshi HARA, National Aerospace Laboratory
viii-lower right	Hiroshi KATO and Seiji GOTO, Tokyo Metropolitan University
ix-upper	Toshio NAGASHIMA, University of Tokyo
ix-lower	Keiichi KARASHIMA, National Aerospace Laboratory
x-lower	Sadatoshi TANEDA, Kyushu University

The Fundamentals of Fluid Flow

1–6, 11, 12	Sadatoshi TANEDA, Kyushu University
7–10, 13, 14, 16	Yasuki NAKAYAMA, Tokai University
15	Taku NAGATA, Gifu University
17–19	Tsuneyo ANDO, Keio University

Laminar and Turbulent Flows

20, 21	Yasuki NAKAYAMA, Tokai University
22	Taku NAGATA, Gifu University
23–26	Jiro SAKAGAMI and Masako MOCHIZUKI, Ochanomizu University
27–29	Sadatoshi TANEDA, Kyushu University
30, 31	Hikaru ITO, Meiji University
32, 33	Tetsuo TAGORI, University of Tokyo
34, 35	Nobuhide KASAGI, Yoichiro IRITANI and Ken HIRATA, University of Tokyo

Laws of Similarity

36–37	Department of Aerodynamic Design, Fuji Heavy Industries Ltd.

38, 39 Takao INUI and Tetsuo TAGORI, University of Tokyo
40 Taku TANAKA, National Ship Laboratory
41 Takeshi MARUO, Yokohama National University

Compressible Flow

42–45 Eisuke OTA and Takamasa OCHIAI, Waseda University
46–48, 50–53 Kazuyasu MATSUO, Kyushu University
49 Yasuki NAKAYAMA, Tokai University, Eisuke OTA, Waseda University
54–59 Keiichi KARASHIMA, National Aerospace Laboratory
60, 61 Takehiko NAGAYAMA, Mitsubishi Heavy Industries Ltd.
62 Yasuo SAITO and Masanobu HONJO, Mitsubishi Heavy Industries Ltd.
63 Takashi IKEDA and Takamasa MITSUZUKA, Toshiba Co. Ltd.
64 Kivohiro TAJIMA, Waseda University

Jets

65, 66 Yasuki NAKAYAMA, Tokai University
67 Shiro AKAIKE, Kanagawa Institute of Technology
68 Yasuki NAKAYAMA, Tokai University, Mitsuzi HAYASHI and Hideo ENDO, National Railroad Laboratory
69 Nobutoshi HARA, National Aerospace Laboratory
70 Kaoru TORII, Yokohama National University
71, 72 Noboru NAKATANI, Osaka University

Miscellaneous Features of the Flow Past Solid Bodies: Wakes

73–76, 78, 79 Toshio KOBAYASHI, University of Tokyo
77, 85 Sadatoshi TANEDA, Kyushu University
80–84 Hidefumi SHIMIZU, Tokyo Institute of Technology
86, 87 Hiroshi KATO and Seiji GOTO, Tokyo Metropolitan University
88 Taku NAGATA, Gifu University
89 Yasuki NAKAYAMA, Tokai University
90, 91, 97 Tetsuo TAGORI, University of Tokyo
92 Shigeru HAYASHIDA, Nagasaki Total University of Technology
93, 94 Taketoshi OKUNO, University of Osaka Prefecture
95, 96 Michitoshi TAKAGI, Nissan Co., Ltd.
98, 99 Shuzo MURAKAMI, University of Tokyo

Pipe and Channel Flows

100–111, 116, 117 Yasuki NAKAYAMA, Tokai University
112–115 Masanobu YAMAMASU, Kanto Gakuin University

Circulation

118–121 Hiroshi NAKAGUCHI, Emeritus Professor, University of Tokyo, Takeichiro HIROSE, University of Tokyo

Wings

122–144, 157	Hikaru ITO, Meiji University
145–149	Susumu MITSUBORI, Takenori NISHI, Masataka SHIRAI and Nobuhiko KAMIYA, National Aerospace Laboratory
150–154	Takenori NISHI, Toshiyoshi FUJITA and Nobuhiko KAMIYA, National Aerospace Laboratory
155–156	Toshio KURIHARA, Tokyo Metropolitan Aero-Technical College
158, 160	Hiroshi NAKAGUCHI, Emeritus Professor, University of Tokyo
159, 163, 164	Hitoshi MURAI, Hiroshima Institute of Technology
161	Sadatoshi TANEDA, Kyushu University
162	Yoshimasa TOMONARI, Nippon Bunri University

Cascades and Fluid Dynamic Machinery

165	Toyoaki YOSHIDA and Fumio MIMURA, National Aerospace Laboratory
166	Kimio SAKATA and Shigemi SHINDO, National Aerospace Laboratory
167, 168	Masanobu YAMAMASU, Kanto Gakuin University
169–171	Junichi KUROKAWA, Yokohama National University
172–187	Takehiko NAGAYAMA, Mitsubishi Heavy Industries, Ltd.
188, 189	Tomitaro TOYOKURA, Yokohama National University
190–192	Yasuki NAKAYAMA, Tokai University

Unsteady Flow

193, 195	Hideo YAMADA, Nagoya Institute of Technology
194	Sadatoshi TANEDA, Kyushu University
196–201	Kiyohiro TAJIMA, Waseda University
202, 203	Tetsuo TAGORI, University of Tokyo

Cavitation

204–208	Hitoshi MURAI, Hiroshima Institute of Technology
209	Shigeo FUJIKAWA and Hideaki AKAMATSU, Kyoto University
210–218	Masao OSHIMA, Ebara Co., Ltd.
219	Gyou YAMAZAKI, Kochi University
220	Hitoshi MURAI, Hiroshima Institute of Technology
221	Seiichi SAITO, Tohoku Gakuin University
222	Hajime YUASA, Mitsui Engineering & Shipbuilding Co., Ltd.
223, 224	Hideo FURUSHIRO, Mitsubishi Heavy Industries, Ltd.
225, 226	Japan Society of Shipbuilding

Behaviour of a Non-Newtonian Fluid

227–231	Hiroshi KATO and Seiji GOTO, Tokyo Metropolitan University
232–233	Norio SHINPO, Tokyo Aerodynamic College
234, 235	Tetsuo TAGORI, University of Tokyo

Summary of Flow Visualization Methods

PICTURES of flow patterns obtained by visualization techniques vary with the state of the flow, its steadiness, the relative motion between the flow and observer, and the method used for flow visualization. In order to help the reader to understand the pictures in this book, the flow visualization methods used to obtain them are outlined below.

Steady flow and unsteady flow

Time-independent and time-dependent flows are called *steady flow* and *unsteady flow* respectively. Whether the flow is steady or not, the pattern observed depends on the relative motion between the flow and the observer. For example, the flow past a body moving at constant velocity in a stationary fluid appears as unsteady flow to an observer who is also stationary, but steady to an observer moving with the body.

Streamlines, path lines, streak lines and time lines

A line everywhere tangential to the velocity of fluid elements at a certain time, is called a *streamline*. The curve which shows where a given fluid element has been at earlier times is called a *path line*. A streamline is a Eulerian description of flow, and a path line is Lagrangian. Figure (i) shows the two-dimensional flow of an ideal fluid past a cylinder. Figure (i)a shows the pattern of streamlines for an observer fixed relative to the cylinder: here the streamlines coincide with the path lines. Figure (i)b shows the pattern of streamlines for an observer travelling with the uniform flow. Figure (i)c shows the path lines for an observer travelling with the uniform flow. In two-dimensional flow, the space between two streamlines is inversely proportional to the local velocity in the flow field.

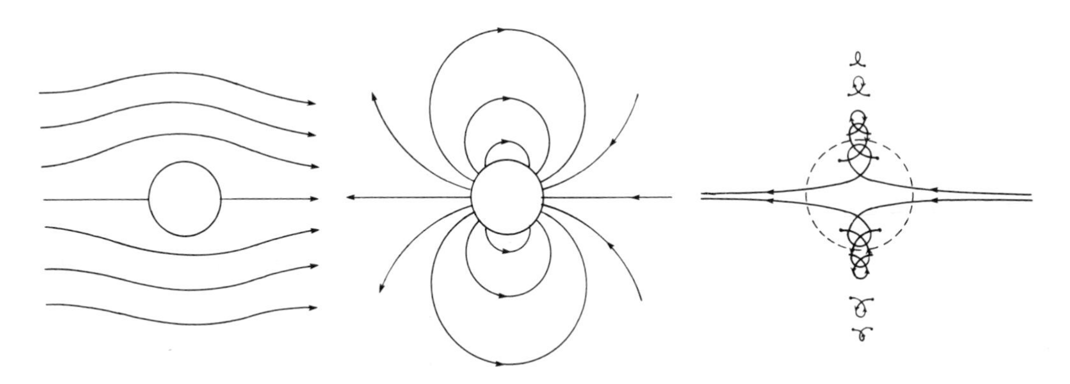

(a) Streamline: still cylinder in uniform flow.

(b) Streamline: cylinder moved in still fluid.

(c) Path line: view from still fluid.

Fig. (i)

A line obtained by steady injection of a fluid tracer from some constant position in the flow, is called a *streak line*. In unsteady flow, a streak line does not coincide with a streamline or a path line.

Figure (ii) shows the pattern of smoke ejected from a chimney over which a wind is blowing. (a) shows the pattern after a time which the wind has remained steady. The wind direction then changes and after a further period of time the smoke pattern appears as shown in (b). The wind direction then changes again and after a further period the pattern appears as shown in (c).

Streamlines, path lines and streak lines do not generally coincide with each other as this example illustrates, although they all coincide in the case of steady flow. Lines obtained by instantaneous injection of a tracer from a source located along a line transverse to the mean-flow, are called *time lines*. Such lines are useful in revealing velocity distributions and flow fluctuations. A line on which physical quantities, for example temperature, pressure or density, etc. maintain a constant value is called a *niveau line* or a *contour line*. There are many methods by which these lines may be obtained.

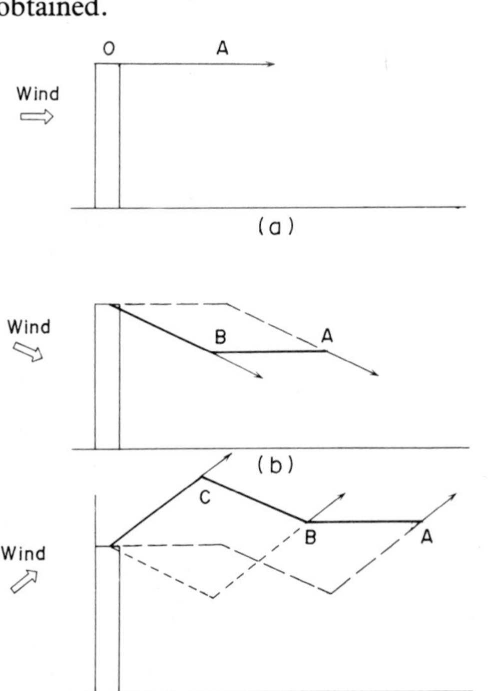

Fig. (ii)

Classification of flow visualization methods

There are many methods of flow visualization and a large number of classifications of them. They have been classified by the flow characteristic quantity (e.g. streamline, path line or streak line) or by the flow phenomenon which they reveal, or by the physical or chemical method used to obtain the visualization: the quality or shape of materials used and the methods for injection or generation of tracers have also been employed as categories, but it is difficult to classify the methods systematically in detail by any single item or principle. Table 1 shows one example of a classification.

Summary of flow visualization methods

(1) Oil film method

A mixture of oil and pigment is spread on the surface of the test model. The shearing force in the flow acts on this oil film, and a striped pattern develops in the deposited pigment. This pattern shows the flow direction on the surface of the model, and can reveal transition, separation and secondary flow patterns. The oil film method is suitable for observing steady flow, and shows the time-averaged flow pattern in unsteady flow. The formation of the pattern is affected by centrifugal force and by gravity.

(2) Temperature sensitive film method

A film of a special material that changes colour or phase according to its temperature, is formed on the surface of the test model. The state of the flow near the model surface is indicated by the colour or phase of this film. Liquid crystal material or temperature sensitive paint is used for this method.

(3) Surface tufts method

Many tufts are attached to the surface of the test model. These indicate the flow direction near the model surface. Separated regions and secondary flows may be detected by the direction and fluctuation of the tufts.

(4) Depth tufts method

Tufts are attached some distance from the

Table 1. Classification of Flow Visualization Method

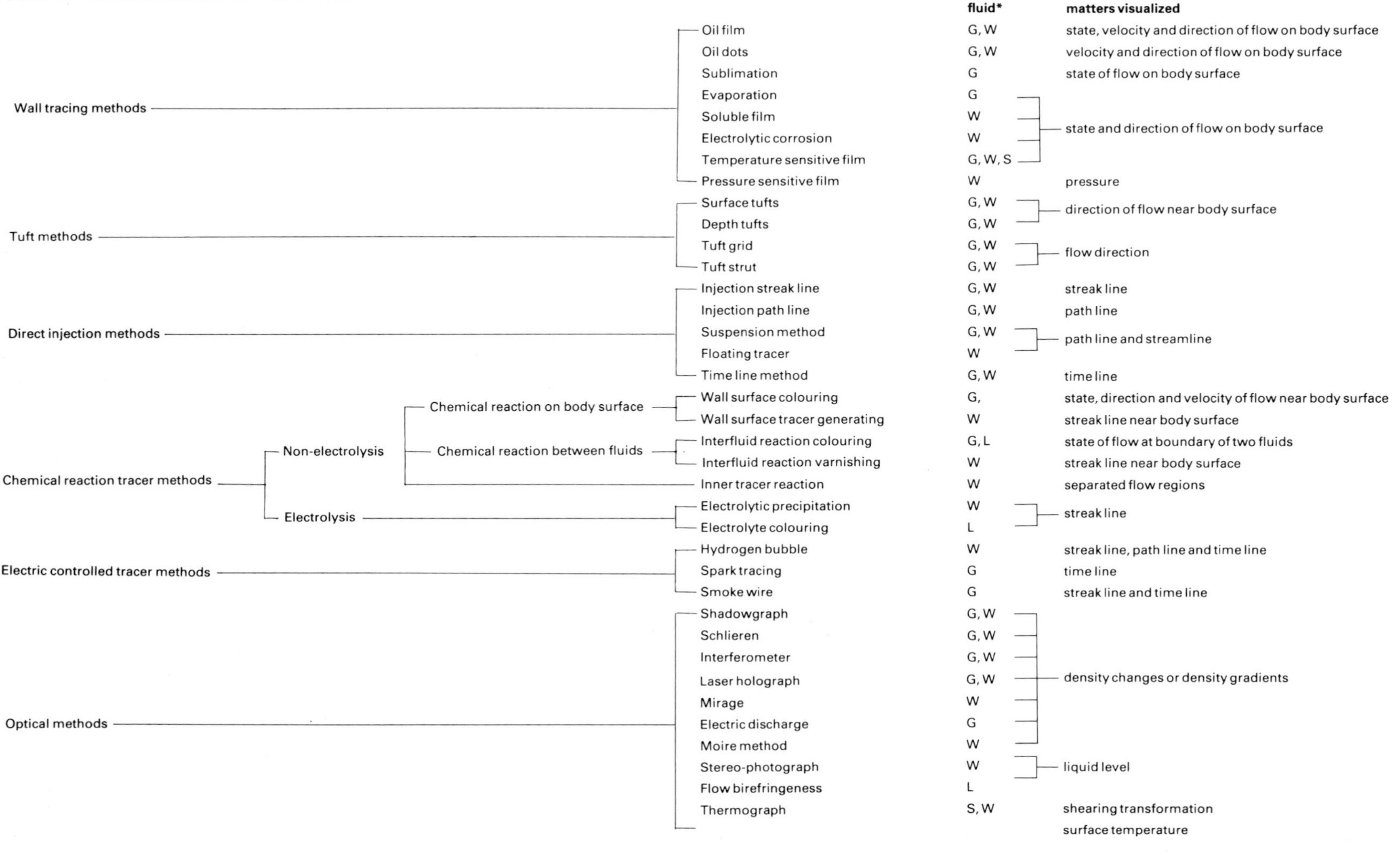

Method group	Subgroup	Reaction type	Method	fluid*	matters visualized
Wall tracing methods			Oil film	G, W	state, velocity and direction of flow on body surface
			Oil dots	G, W	velocity and direction of flow on body surface
			Sublimation	G	state of flow on body surface
			Evaporation	G	
			Soluble film	W	state and direction of flow on body surface
			Electrolytic corrosion	W	
			Temperature sensitive film	G, W, S	
			Pressure sensitive film	W	pressure
Tuft methods			Surface tufts	G, W	direction of flow near body surface
			Depth tufts	G, W	
			Tuft grid	G, W	flow direction
			Tuft strut	G, W	
Direct injection methods			Injection streak line	G, W	streak line
			Injection path line	G, W	path line
			Suspension method	G, W	path line and streamline
			Floating tracer	W	
			Time line method	G, W	time line
Chemical reaction tracer methods	Non-electrolysis	Chemical reaction on body surface	Wall surface colouring	G,	state, direction and velocity of flow near body surface
			Wall surface tracer generating	W	streak line near body surface
		Chemical reaction between fluids	Interfluid reaction colouring	G, L	state of flow at boundary of two fluids
			Interfluid reaction varnishing	W	streak line near body surface
			Inner tracer reaction	W	separated flow regions
	Electrolysis		Electrolytic precipitation	W	streak line
			Electrolyte colouring	L	
Electric controlled tracer methods			Hydrogen bubble	W	streak line, path line and time line
			Spark tracing	G	time line
			Smoke wire	G	streak line and time line
Optical methods			Shadowgraph	G, W	
			Schlieren	G, W	
			Interferometer	G, W	
			Laser holograph	G, W	density changes or density gradients
			Mirage	W	
			Electric discharge	G	
			Moire method	W	
			Stereo-photograph	W	liquid level
			Flow birefringeness	L	
			Thermograph	S, W	shearing transformation
					surface temperature

* G: gas, W: water, L: liquid, S: solid

surface of the test model on fine struts. The direction of flow past the model and features such as vortices may be detected by the direction of these tufts. Sometimes surface tufts are used together with depth tufts.

(5) Tuft grid method

A large number of tufts are located on a mesh network in a plane usually at right angles to the flow. The flow around the test model, its wake and any trailing vortices are revealed by the pattern of the tufts.

(6) Injection streak line method

Fluid tracer is injected continuously into the flow from fine tubes, or from holes or a slot on the test model. The state of flow is shown by the resulting streak lines.

(7) Suspension method

Particles of fluid or solid tracer are distributed uniformly in the fluid, usually before movement is initiated. The state of flow is revealed by the paths of the tracer particles.

(8) Floating tracer method

The state of flow is examined by observing the motion of tracer particles that are dispersed and floated on the free surface of a liquid. This method also yields path lines. It must be borne in mind that the free surface flow may differ from that below the surface.

(9) Time line method

The fluid tracer is injected to the flow in line, generally in a transverse straight line. This line of tracer is used as a time line, and the nature of the flow is investigated by observing the time transformation of this time line.

(10) Electrolytic precipitation method

This method is applicable when water is used as the working fluid. The model has areas of metal plating on its surface or a fine wire is placed in the upper stream of the model. These act as an anode. White particles of precipitate are generated near this anode by electrolysis, and these act as tracers. Tin or tin alloy is usually used as the anode material. This method is a kind of electrolytic precipitation method.

(11) Hydrogen bubble method

In water flows, small hydrogen bubbles are generated by electrolysis on a fine wire used as a cathode. These bubbles act as the tracer. Streak lines are formed by using a steady voltage and a wire kinked to a zig-zag shape or insulated at intervals. Time lines are generated by the use of a straight wire and voltage pulses.

(12) Spark tracing method

Electric discharges are produced using a series of high voltage pulses between suitably shaped electrodes. The first electric spark makes an ionized path. This ionized path moves together with the air flow and the second electric spark propagates along this moving path which has very low electric resistance. In such a manner, the subsequent electric sparks propagate along the moving ionized path one after another and trace time lines of the air flow.

(13) Smoke wire method

White smoke is generated from a metallic wire spread with oil and heated by a pulse of electric current. The smoke so generated is used as the tracer to visualize the flow.

(14) Shadowgraph method

The light of a point source or a parallel beam of light is passed through the flow in a test section and projected on to a screen or a photographic film. The state of the flow is indicated by the pattern of brightness. The method is sensitive to the second spatial derivative of the fluid density (Figs. (iii) and (iv)).

(15) Schlieren method

A parallel beam of light is refracted through the flow, focused on to a knife edge, and then used to form an image on the screen

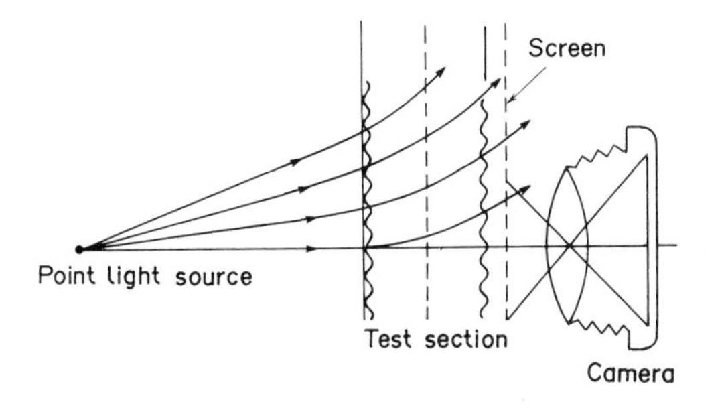

Fig. (iii)

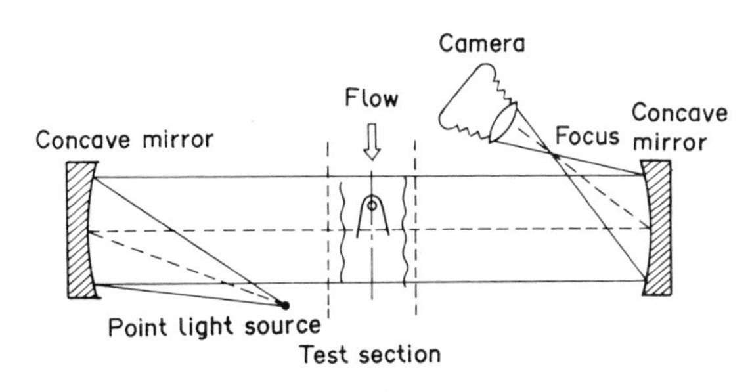

Fig. (iv)

or photographic film behind the knife edge. The state of flow is visualized by a bright and dark pattern of the screen or film. This method is sensitive to the first spatial derivative of the fluid density (Fig. (v)).

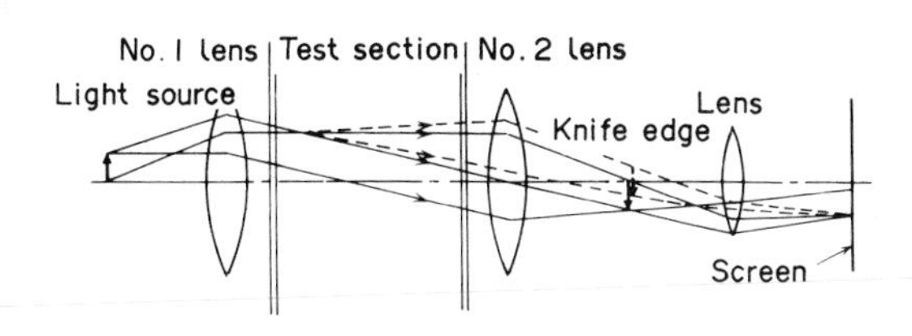

Fig. (v)

(16) Interferometer method

A parallel beam of light is split and one of the beams is passed through the flow in the test section, then recombined with the other beam, which has passed through a compensator section. The combined beams then form an image on a screen or photographic film. The flow is visualized by the resulting interference fringe pattern. Equi-density lines of the flow field are obtained by adjusting the system to give uniform illumination over the screen with still air in the working section. Transformed fringes are obtained in proportion to the density change of flow by adjusting the system to give parallel fringes on the screen with the fluid in the test section at rest (Fig. (vi)).

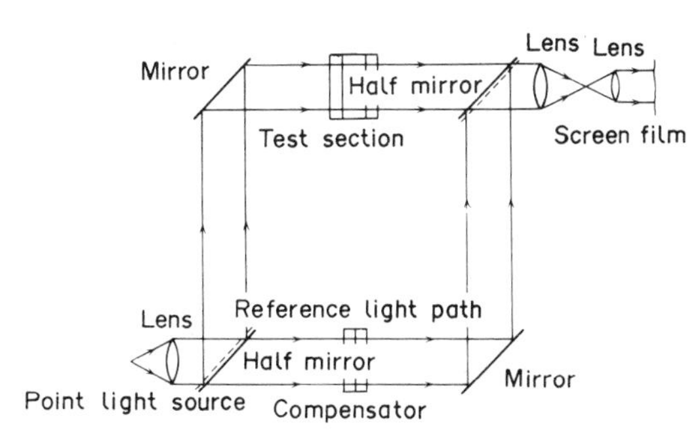

Fig. (vi)

(17) Thermograph method

The infra red rays radiated from the surface of a solid body or a liquid are detected and the distribution of surface temperature is visualized as a bright and dark pattern of equi-temperature lines.

(18) Hele-Shaw flow method

A two-dimensional model is installed in the small clearance between two parallel flat glass plates. The flow of a viscous fluid in this narrow space has streamlines which coincide with those of potential flow. Accordingly, the streamlines of two-dimensional potential flow are obtained by continuous injection of a dye or hydrogen bubbles centrally between the plates.

(19) Shallow water tank method

When the water depth is shallow enough, the equation of motion of open channel flow is approximately analogous to the equation of gas dynamics with a value of the isentropic index of 2. Shock waves generated in the gas dynamics case, correspond to the water waves. The sudden increase in density is equivalent to the almost discontinuous rise in the height of the free surface of the shallow water. The flow of a compressible fluid may be investigated by the observation of the wave pattern.

The Fundamentals of Fluid Flow

Stokes' flow

The flow pattern around the body varies with the *Reynolds number*. When the Reynolds number is much smaller than unity, the inertia forces are negligible compared with the viscous forces and the flow can be described by Stokes' approximation. In a Stokes' flow, the streamline pattern around a solid body having a symmetrical shape is symmetrical and resembles superficially the pattern of potential flow. Figure 1 shows Stokes' flow around a circular cylinder in a fluid.

Flows around a circular cylinder

The Stokes' approximation is invalid at Reynolds numbers larger than about unity. Figure 2 shows the flow around a circular cylinder at $Re = 1.1$, where Re is the Reynolds number. *Flow separation* does not occur at a Reynolds number of this order, but the streamline pattern is no longer symmetrical.

With a further increase of Reynolds number, a pair of vortices is formed at the rear of the circular cylinder as shown in Figs. 3 and 4. The critical Reynolds number at which the *vortex pair* appears is about 6. The length of the vortex pair increases with the Reynolds number in the range $6 < Re < 40$.

When the Reynolds number exceeds about 40, the wake of a circular cylinder begins to develop a progressive wave motion, and the vortex pair breaks down. The vortices leave the cylinder and move downstream as shown in Fig. 5. The critical Reynolds number at which the wake begins to oscillate is affected by external disturbances, but lies between 30 and 45.

Karman vortex street

When a cylindrical body is placed in a uniform flow, the wake performs a progressive wave motion at Reynolds numbers larger than about 40. The wavelength is approximately equal to the wake width. Vortices are shed alternately from each side of the cylinder and persist for some distance downstream, forming a double row called a Karman vortex street. Figure 6 shows the Karman vortex street behind a circular cylinder, made visible by means of the electrolytic precipitation method and the white curves are *streak lines*.

Figures 7–9 also show Karman vortex streets behind circular cylinders: one with time lines in Fig. 7: one with streak lines in Fig. 8: one with both time lines and streak lines together in Fig. 9. It is very convenient that the velocity at a point in the stream can be calculated from the pulse frequency multiplied by the distance between the time lines, and that complex flow direction and magnitude can also be obtained from Fig. 9. Figure 10 shows a Karman vortex street behind a triangular prism. The vortices here are clearer than those formed behind a circular cylinder.

A Karman vortex street frequently forms behind an obstacle in a stream, and the periodic vortex shedding creates a cyclic force on the obstacle. Periodic strong winds in the lee of skyscrapers, oscillation of suspension bridges, and humming of tele-

Fig. 1. Flow around a circular cylinder at $Re = 0.038$ (glycerine, flow velocity 0.15 cm/s, cylinder diameter 1.0 cm, tank width 40 cm, aluminium powder method).

Fig. 2. Flow around a circular cylinder at $Re = 1.1$ (glycerine–water solution, flow velocity 0.20 cm/s, cylinder diameter 1.0 cm, aluminium powder method).

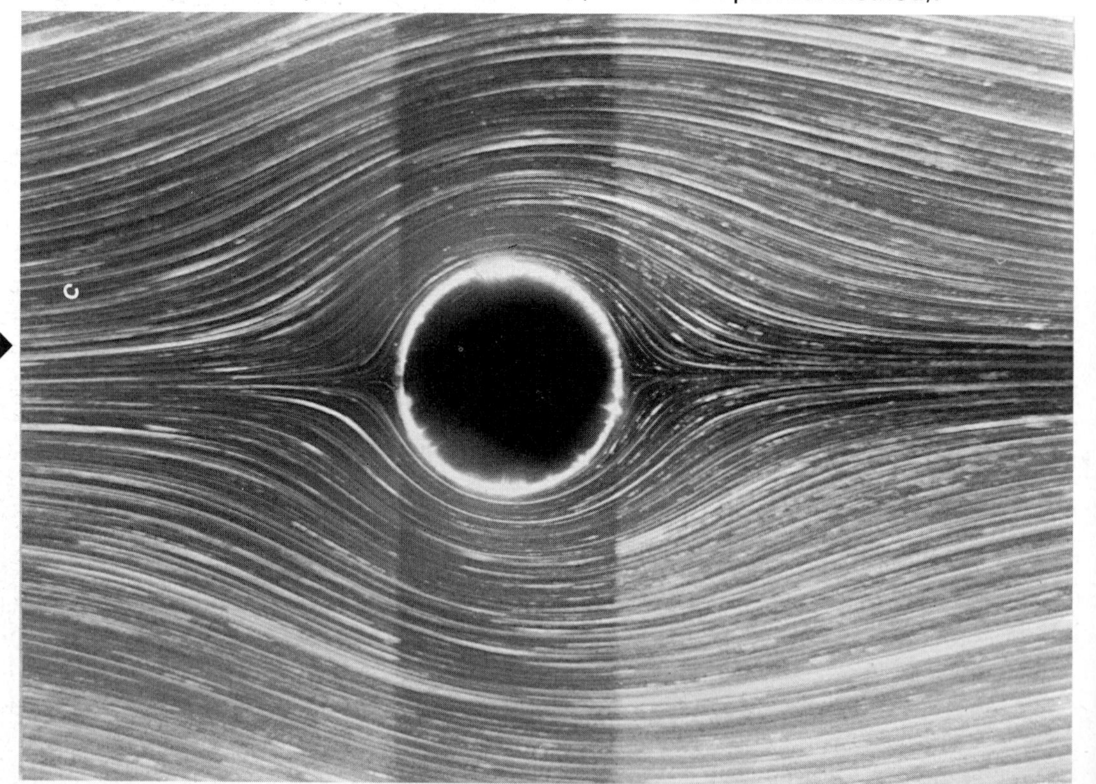

phone wires in wind are all caused by *vortex shedding*.

Because the frequency of a Karman vortex street is proportional to the velocity of the flow, a type of vortex flow meter which measures flow rate by detecting vortices, has been put to practical use. Triangular prisms are often used to create the vortices for such flow measurements.

Couette flow

Figure 11 shows the velocity distribution in a viscous fluid between two parallel plates. With the two plates at rest, the space between the two plates was filled with glycerine, and a syringe was used to inject a small quantity of red glycerine in such a way as to make a straight line perpendicular to the plates. The upper plate was then moved in its own plane at a constant speed while the lower one was held stationary. The resulting deformation of the red line was then photographed. The flow generated in this way without any pressure difference is called a Couette flow.

Two-dimensional Poiseuille flow

Figure 12 also shows the velocity distribution in a viscous fluid between two parallel flat plates. In this case, however, the two plates are fixed and the glycerine between the two plates is moved by applying a pressure difference. This type of flow is called Poiseuille flow. Two-dimensional Poiseuille flow remains laminar when the Reynolds number is smaller than about 3000.

Fig. 3. Flow around a circular cylinder at Re = 19 (water, flow velocity 0.20 cm/s, cylinder diameter 1.0 cm, aluminium powder method and electrolytic precipitation method).

Fig. 4. Flow around a circular cylinder at Re = 26 (water, flow velocity 0.25 cm/s, cylinder diameter 1.0 cm, aluminium powder method).

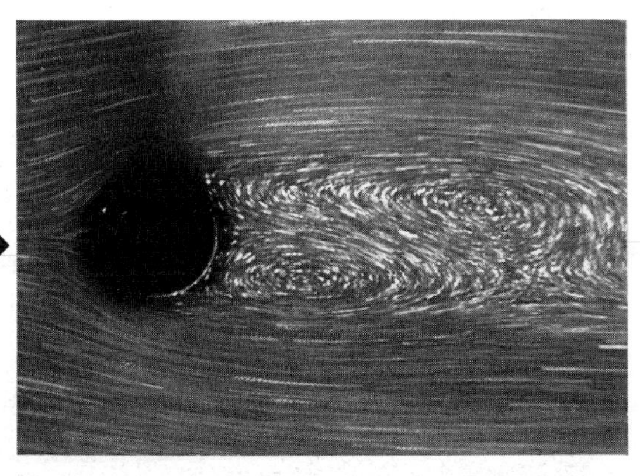

Fig. 5. Flow around a circular cylinder at Re = 55 (water, flow velocity 0.55 cm/s, cylinder diameter 1.0 cm, aluminium powder method).

(The Reynolds number here is Uh/v, where U is the mean velocity, h the distance between the two plates, and v is the kinematic viscosity of the fluid.)

Figure 14 shows the same flow with a series of time lines originating from six stations down the pipe. The development of the parabolic velocity profile may be seen very clearly.

Flow in an inlet region and Poiseuille flow

Figure 13 shows flow entering a circular pipe with a well designed short *bell mouth* inlet. The velocity of fluid particles in contact with the wall is zero, and the velocity in the layers near the wall is decreased by the action of viscosity. The region of reduced flow velocity becomes progressively thicker with distance downstream.

Finally, the velocity distribution becomes parabolic. The distance from the inlet to the pipe to this point is called the inlet region. The flow beyond the inlet region, where the velocity distribution is parabolic, is another example of Poiseuille flow.

Separation

When water is electrolyzed, using as a positive terminal a circular cylinder whose surface is covered by tin alloy, colouring matter is produced from the tin alloy surface and settles on it. Fluid particles near the circular cylinder are coloured by this precipitation. When a sudden impulse is given to a circular cylinder, separation of the flow takes place from the *boundary layer* and a separation vortex is formed. Figure 15 show an instance of such a flow separation formation. Figure 15 also shows that a pair of secondary vortices is formed near the separation point in addition to the large vortex which is shed.

Fig. 6. Karman vortex street behind a circular cylinder at $Re = 140$ (water, flow velocity 1.5 cm/s, cylinder diameter 1.0 cm, electrolytic precipitation method).

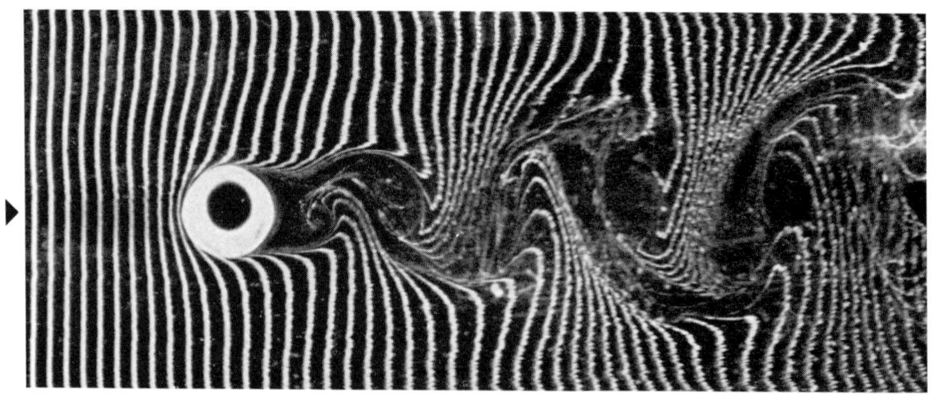

Fig. 7. Time lines.

Fig. 8. Streak lines.

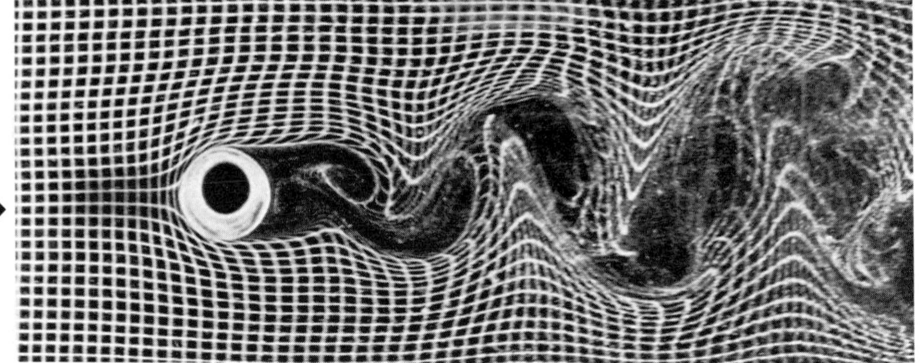

Fig. 9. Time lines and streak lines.

Figs. 7–9. Karman vortex street behind a circular cylinder at $Re = 170$ (water, flow velocity 2.6 cm/s, cylinder diameter 8 mm, hydrogen bubble method).

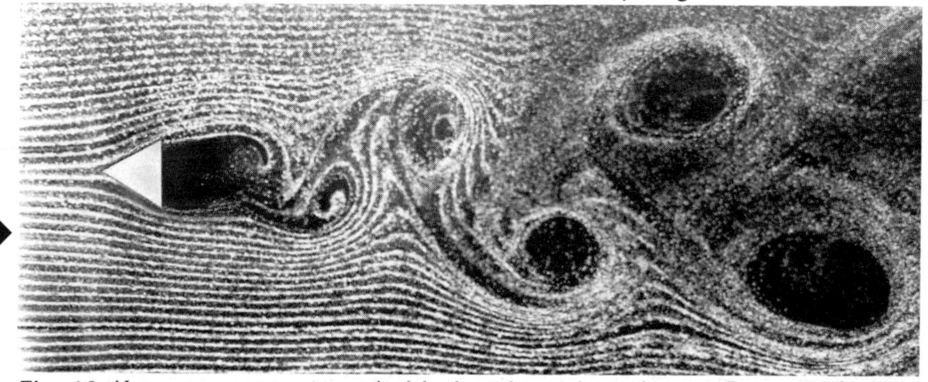

Fig. 10. Karman vortex street behind a triangular prism at $Re = 170$ (water, flow velocity 2.6 cm/s, length of triangle side 7 mm, hydrogen bubble method).

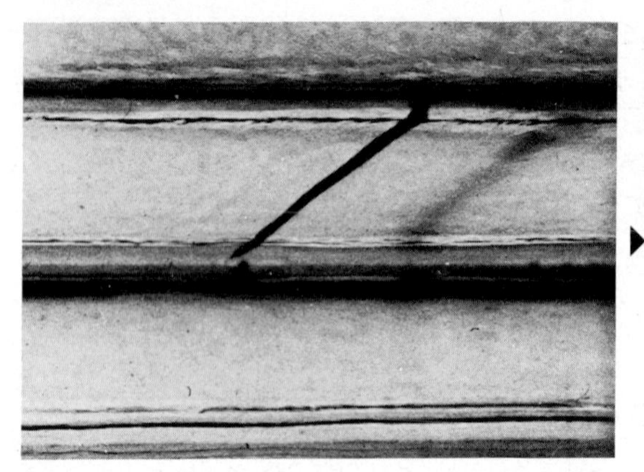

Fig. 11. Couette flow (glycerine, upper plate speed 2 mm/s, distance between two plates 20 mm, $Re = 2.7 \times 10^{-2}$, coloured line method).

Fig. 12. Two-dimensional Poiseuille flow (glycerine, centre velocity 2 mm/s, distance between two plates 40 mm, $Re = 5.3 \times 10^{-2}$, coloured-line method).

Deformation and rotation

In general, fluid motion is made up of a combination of *translation*, rotation and deformation.

Figure 16 shows the deformation, translation and rotation of fluid particles in an asymmetric contracting channel flow, the flow being made visible by the hydrogen bubble method. The bubbles are formed by a partially insulated tungsten filament used as an electrode.

Fluid particles near the centre of a convergent duct are deformed into a long and narrow shape by the longitudinal acceleration of the fluid. Fluid elements near the wall are deformed into a distorted, sheared and rotated shape, due to the velocity gradient in the transverse direction.

Hele-Shaw flow

When the distance between the walls is sufficiently small, the slow flow of a viscous liquid between two parallel walls has properties analogous to a potential flow. This type of flow is known as Hele-Shaw flow.

To make the flow visible, dyed liquid is used as a tracer, and the walls are transparent glass plates. Figures 17–19 show flows passing around circular and elliptic cylinders and a symmetrical *Joukowski aerofoil*. The flow is steady and the *streak lines* are also streamlines.

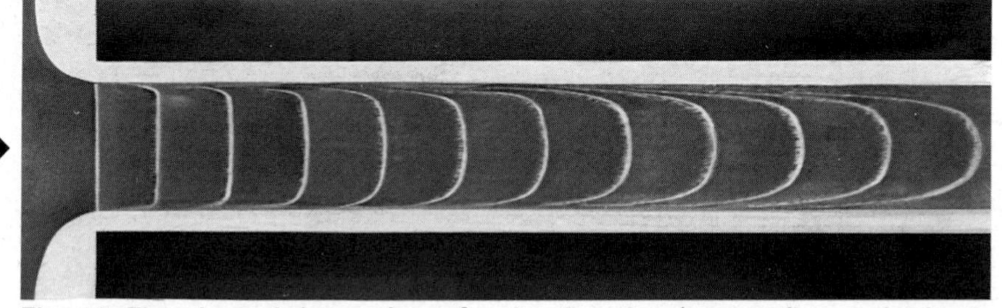

Fig. 13. Flow in an inlet region of a circular pipe (water, flow velocity 6 cm/s, pipe diameter 27 mm, $Re = 1.6 \times 10^3$, hydrogen bubble method).

Fig. 14. Flow in an inlet region between parallel plates (water, flow velocity 3.2 cm/s, distance between two plates 20 mm, $Re = 640$, hydrogen bubble method).

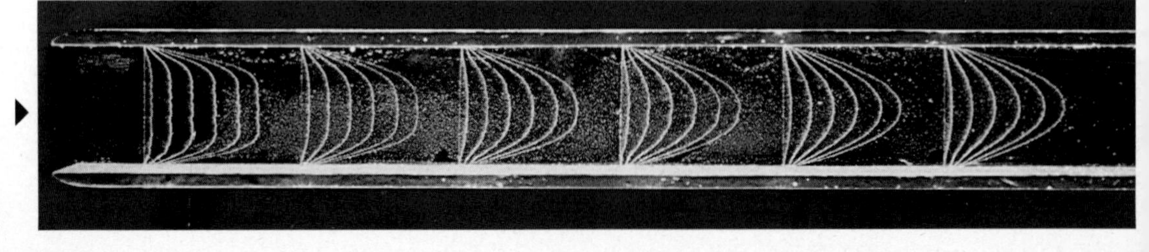

Fig. 15. Separation of flow around a circular cylinder started impulsively (water, flow velocity 3.2 cm/s, cylinder diameter 70 mm, $Re = 2 \times 10^3$, lapse of time from start 4.2 s, electrolytic precipitation method).

Fig. 16. Deformation and rotation of fluid particles in an asymmetric contraction channel flow (water, inlet velocity 2.8 cm/s, inlet width 90 mm, outlet width 23 mm, $Re = 2.5 \times 10^3$, hydrogen bubble method).

Hele-Shaw flow is realized when the velocity of flow is sufficiently small for inertia forces to be negligible. If the reference length of the body is denoted by L, the clearance by h, and the velocity of the main flow by U, then the Reynolds number is limited by $Re = (UL/v)(h/L)^2 << 1$.

Fig. 17. Slow flow passes a circular cylinder (cylinder diameter 8 cm).

Fig. 18. Slow flow passes an elliptic cylinder (length of the major axis 12 cm, length of the minor axis 6.2 cm).

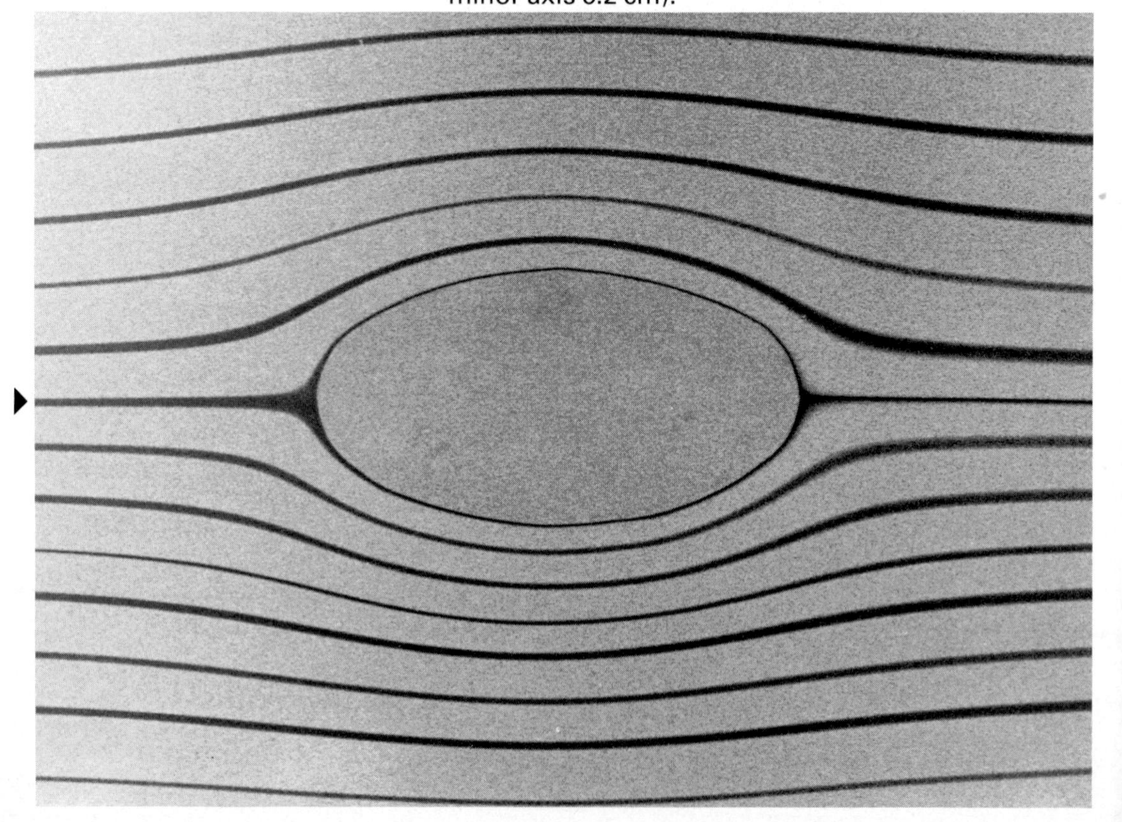

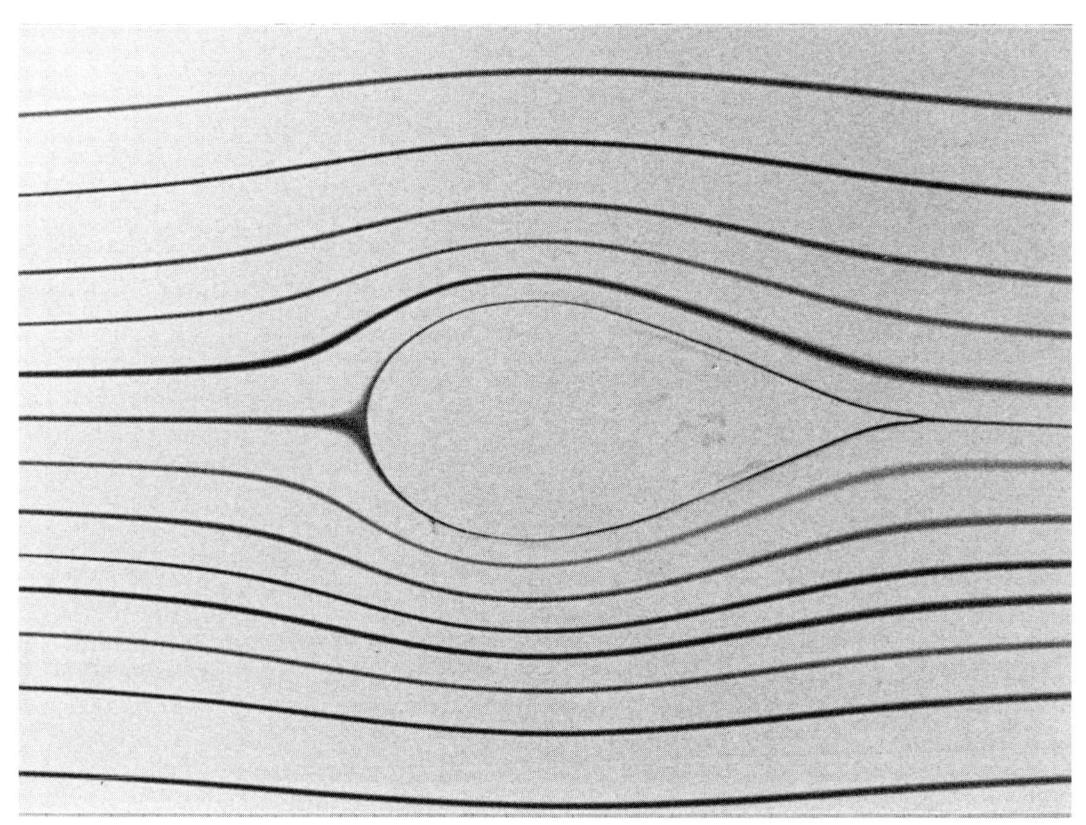

Fig. 19. Slow flow passes a symmetrical Joukowski aerofoil (chord length 14 cm).

Figs. 17–19. Water, flow velocity 2.8 cm/s, injection streak line method.

Laminar and Turbulent Flows

Laminar boundary layer along a flat plate

Consider a solid body placed in a uniform flow of a fluid of low viscosity. The fluid immediately adjacent to the body surface is at rest relative to the body for most cases except for gases under conditions of very low pressure, while in the flow field far from the body surface the influence of the viscosity is not felt and the flow exhibits no shear deformation. The influence of viscosity is confined to a thin region close to the body, where the transverse velocity gradient is large, and though the viscosity is small, the shear stress becomes significant and the fluid velocity continuously changes from zero on the surface to the uniform flow velocity. Thus the body is covered with a thin layer, called the *boundary layer*, characterized by a large transverse velocity gradient.

When a semi-infinite flat plate is placed parallel to the flow direction in a uniform flow of velocity U, a boundary layer develops along it. The thickness δ of this boundary layer gradually increases in the streamwise direction, and the flow within it usually remains laminar until the *Reynolds number*, $Re = Ul/\nu$, reaches an order of 10^5, where l is the streamwise distance from the leading edge of the plate, and ν is the kinematic viscosity of the fluid.

In Fig. 20, the development of the laminar boundary layer along a flat plate is visualized by the *hydrogen bubble method*. A fine electrode wire is introduced upstream of the flat plate and voltage pulse is applied repeatedly at regular intervals. The *boundary layer thickness* is seen to increase with the streamwise distance.

The velocity profile in a laminar boundary layer is visualized by the use of a electrode wire normal to the surface in Fig. 21. The velocity profile has also been obtained numerically and is known as the *Blasius profile*.

Unsteady boundary layer around an impulsively starting cylinder

When a flow separation occurs in the *steady* flow around an unstreamlined bluff body, the flow pattern observed is very different from that deduced by the *potential flow* theory. Downstream of the *separation* point, a large vortical flow region appears accompanied by flow reversal near the body surface.

When a circular cylinder starts impulsively at a constant velocity in a stationary surrounding fluid, the initial flow pattern is that predicted by potential flow theory, but a boundary layer develops very rapidly and reversed flow close to the surface soon appears. Subsequently separation vortices are formed and the periodic vortex shedding regime illustrated in Figs. 7–9 may be established. Figure 22 shows *time lines* at constant intervals produced by the *hydrogen bubble method* with eight electrode wires normal to the cylinder surface. Each relative distance between the timeline and the electrode wire corresponds to the instantaneous flow velocity near the electrode. The flow reversal in the boundary layer can be seen clearly near the surface at the downstream 4th to 8th electrode wires. At the moment this photograph was taken, the *separated region* was still thin enough to ensure that the flow outside the boundary layer could be approximated by the potential flow theory in contrast to steady separated flow, of the type shown in Fig. 3. As shown in Fig. 22, the appearance of a flow reversal in the unsteady boundary layer does not necessarily imply

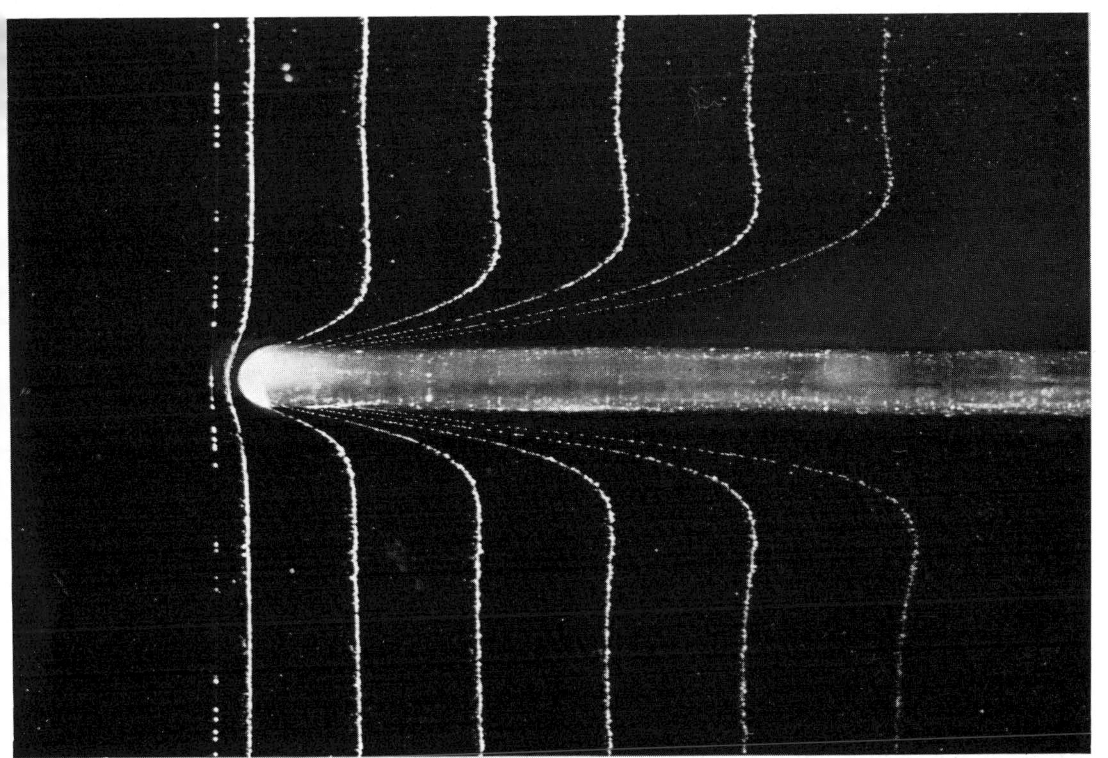

Fig. 20. Development of laminar boundary layer (0.01% salt water, free steam velocity 0.6 cm/s, thickness of the plate 0.5 mm, hydrogen bubble method).

the formation of a large vortical region downstream.

Laminar to turbulent transition: Tollmien-Schlichting (TS) waves

When a two-dimensional sinusoidal wave of very small amplitude is introduced into the laminar boundary layer on a flat plate, the wave is either gradually amplified or eventually damped out as it travels downstream. Whether it is amplified or damped is determined by the wave number and the Reynolds number based on the boundary layer thickness. Spontaneous occurrences of waves of this type, known as TS waves, initiate the laminar to turbulent transition of boundary layer flow.

Figures 23 and 24 show a plan view of a flat plate, over which air is flowing with a free stream velocity of 5 m/s. A metal ribbon of 0.2 mm in thickness and 2 mm in width is positioned parallel to the flat plate in the boundary layer, and is vibrated in a direction perpendicular to the flow. The vibration is maintained by an oscillating magnetic force, induced by a permanent magnet beneath the plate and an AC current through the

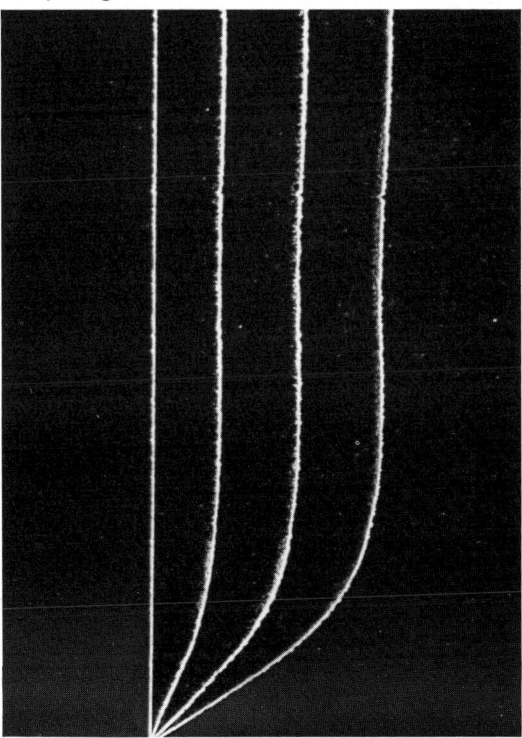

Fig. 21. Velocity profile in the laminar boundary layer (0.01% salt water, free stream velocity 0.6 cm/s, distance from the leading edge 200 mm, $Re = 1.2 \times 10^2$, hydrogen bubble method).

Fig. 22. Water, velocity of motion 2 cm/s, cylinder diameter 70 mm, photographed two seconds after the start of motion, $Re = 1.2 \times 10^3$, hydrogen bubble method.

ribbon. *Smoke film* tracer is introduced 50 cm downstream of the vibrating ribbon where the TS waves have been sufficiently amplified to be visualized. The TS waves thus artificially created may be amplified or damped with streamwise distance in the boundary layer, depending on the frequency and amplitude of the vibration, and also the distance of the vibrating ribbon from the wall. The frequency is 30 Hz and the distance is 1 mm, while the electric current in the ribbon is changed from 0.5 to 1, 2 and 3.5A from the top to the bottom frame in Fig. 23. In all cases the TS waves are being amplified. The waves are initially perpendicular to the flow, but as the amplification process proceeds, three-dimensional distortion appears and finally results in transition to turbulence.

Figure 24 is a close-up of the distortion of the TS waves. These photographs were taken stereographically (one of a pair of stereographs is shown here), so that the smoke pattern could be observed three-dimensionally. The result suggests that the plain two-dimensional TS waves, initially perpendicular to the flow, eventually roll up and develops the spanwise irregularity visible in Fig. 24. Sections of these transverse rolls lift up into the outer flow and are swept downstream, trailing behind them diagonal scrolls in which tracer streaks can be seen twisted round each other as in a vortex filament.

The turbulent spot

The set of photographs in Fig. 25 are sequential plan views of the boundary layer along a flat plate taken at intervals of about 0.8 s. The layer of smoke was introduced uniformly into the boundary layer some distance upstream. The deformation of the *smoke layer* visible in Fig. 25 was triggered upstream by a sudden projection of a small circular cylinder into the boundary layer. The diameter of the cylinder was 1.6 mm and it was projected through a small hole into the boundary layer 1.6 mm above the flat plate surface for about 0.6 s. The upper

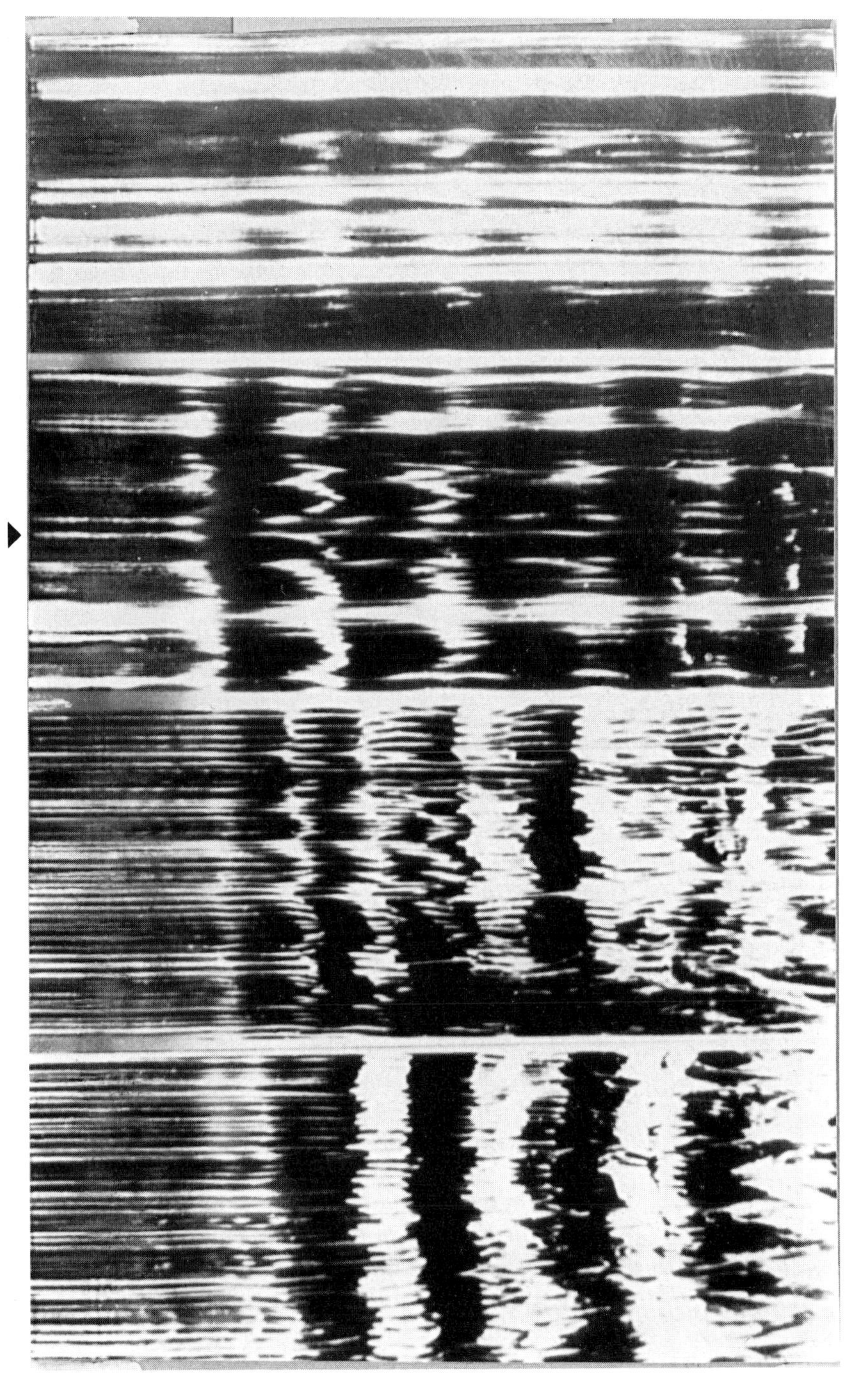

Fig. 23. Air, free stream velocity 5 m/s, distance of the vibrating ribbon from the wall 1 mm, layer of paraffin smoke.

Fig. 24. Sections of these transverse rolls lift up into the outer flow and are swept downstream, trailing behind them diagonal scrolls in which tracer streaks can be seen twisted around each other as in a vortex filament.

photograph was taken just after the projection of the cylinder, while the rest are taken successively. The wedge of corrugated smoke film pointing downstream, called a *turbulent spot*, originates from the artificial disturbance given by the upstream cylinder. It gradually develops and evenually grows into an element of turbulence. In the lowest photograph taken after the passing of the turbulent spot, the dark part without smoke also shows the shape of a wedge, but pointing upstream. This region is gradually refilled with smoke and eventually becomes covered once more with a uniform smoke layer.

Laminar to turbulent transition triggered by a sphere

Photographs in Fig. 26 are plan views of flow visualization downstream of a sphere attached to a flat plate. The laminar boundary layer along the plate is disturbed by the wake of the sphere, and three-dimensional *horseshoe vortices* slanting downstream are produced. The left-hand side 1/5 of each photograph was taken with smoke emitted from a narrow slit positioned 7 mm downstream of the sphere. The remaining 4/5 of each photograph was obtained using a *smoke layer* produced from a heated nichrome ribbon parallel to the surface 50 mm downstream of the sphere. In Fig. 26, from the top to the bottom, the ribbon was positioned at distances of 2, 6, 8, 10, 11 and 16 mm from the surface respectively. Hence, each photograph shows the part of the flow structure which originates from a different level.

Transition from laminar to turbulent flows

In the boundary layer flow along a flat plate parallel to the flow direction, *laminar to turbulent transition* takes place around the location where the *Reynolds number* based on streamwise distance from the leading edge reaches about 10^5. However, quite weak disturbances in the free stream may cause a considerable reduction in this *critical Reynolds number*. Figure 27 shows transition occurring in the boundary layer along a horizontal flat plate in a wind tunnel. A thin

Fig. 25. Air, flow velocity 4 m/s, $Re = 450$, cylinder diameter 1.6 mm, layer of paraffin smoke.

titanium tetrachloride *smoke layer* has been introduced into the boundary layer from the leading edge. The upper photograph is a plan view, while the lower is a vertical streamwise cross-sectional view. (These photographs were not taken simultaneously.) Transition phenomena are quite unstable and the point of transition moves back and forth continually. Many crescent shaped regions may be seen in the smoke near the laminar-turbulent boundary. These seem to be the active regions of *turbulent spots*. Once transition occurs, the boundary layer thickness increases rapidly and the outer edge of the layer exhibits a complex corrugated shape. Many elaborate studies of transition have been undertaken, but the details of its mechanism still remain open to question.

Taylor vortices

In the fluid circulating between two concentric *rotating cylinders*, flow instability due to *centrifugal force* may occur and this results in secondary ring vortices forming around the inner cylinder. These are called *Taylor vortices*. This name originates from G. I. Taylor who first investigated in detail this kind of flow instability, both theoretically and experimentally. The axial spacing between two neighbouring vortices is almost equal to the gap width between the two cylinders.

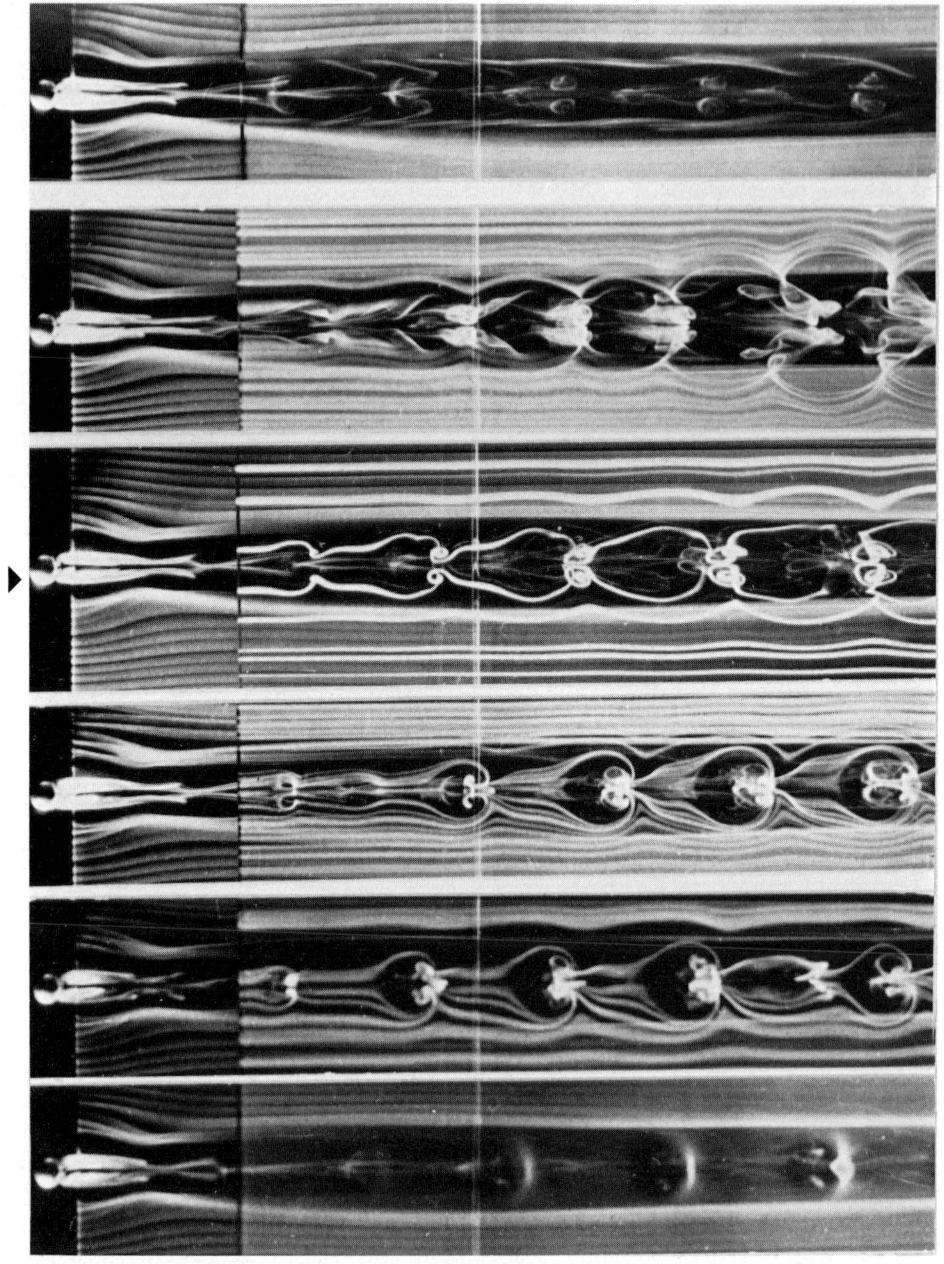

Fig. 26. Air, free stream velocity 1.5 m/s, $Re = 746$, diameter of a sphere 7 mm, paraffin smoke.

Laminar Taylor vortices are visualized by the *aluminium powder* method in Fig. 28. The cylinder gap has been filled with water. The inner cylinder rotates at a relatively low speed, while the outer cylinder is at rest. The tiny aluminium powder particles suspended in the water resemble fish scales in shape, and they tend to align in the local flow direction. Once Taylor vortices occur between the cylinders, then the flow field becomes periodic in the axial direction. Accordingly, the light reflected from the aluminium powder also exhibits an axial periodicity, and this permits the axial wave length of the vortices to be determined.

With increasing rotational speed of the inner cylinder, azimuthal waves are imposed on the ring vortices, and eventually the flow becomes turbulent as shown in Fig. 29. It is interesting to note that the ring vortices still exist even in the turbulent flow regime, and that their axial wavelength is little changed from that of the laminar regime.

Fig. 27. Air, free stream velocity 3.3 m/s, $Re = 2 \times 10^5$ (*Re* based on the mean distance from the leading edge to the transition), plate width 1.2 m, plate length 2.4 m, smoke film method.

Görtler vortices on a concave surface

In the boundary layer along a surface with concave curvature in the direction of flow, an array of streamwise vortices is produced. These are known as *Görtler vortices*. The photographs in Figs. 30 and 31 were taken using strobe lighting with a uniform layer of *light-oil vapor* mist introduced beneath the boundary layer as shown in the schematic figure. The radius of the wall curvature is 1 m, and the free stream velocity is 2.5 m/s. A general view of the flow is shown in Fig. 30, where numbers denote the distance along the concave surface from the leading edge. In the upstream region, the vapor concentrates into low-speed regions between pairs of counter-rotating *streamwise vortices*. Downstream of a line about 850 mm from the leading edge, three-dimensional horseshoe-like vortices appear riding on the pairs of streamwise vortices. The *horseshoe vortices* eventually break down at a position about 1000 mm from the leading edge, and turbulent boundary layer is established downstream.

The sequence above is more clearly studied by visualizing the flow field in a plane perpendicular to the flow direction as shown in Fig. 31. The upwelling motion associated with the streamwise vortices is seen in mushroom shape of tracer, of which the inner part corresponds to the low-speed fluid flowing from the region between a pair of vortices in the upstream boundary layer. The horseshoe vortices, mentioned above, are formed along the upper boundary between the streamwise vortex pair and the free stream. The rotational motion of the streamwise vortices is weakened by the development of the horseshoe vortices, and transition from a laminar to a turbulent boundary layer results.

Transition in a pipe flow

In the flow in a circular pipe, the fluid moves in ordered layers when the *Reynolds number* is small enough, but irregular mixing of fluid begins to take place when the Reynolds number exceeds about 2300. At

Reynolds numbers just exceeding the critical value, the mixing takes place in irregularly spaced bursts, but at Reynolds numbers higher still, the mixing is continuous and apparently random. These stages are called *laminar*, *transitional* and *turbulent* flow respectively. In Fig. 32, dye is introduced into the flow in a circular pipe from an upstream fine tube. When the Reynolds number is small, so that the flow is kept laminar, the streak line does not diffuse. But, as the Reynolds number is increased, the *streak line* shows chaotic mixing due to the development of turbulence.

Flow on a rotating disk

Using the *oil film method*, a spiral pattern is formed on a rotating disk, since the fluid near the disk surface is given a radial velocity component by *centrifugal force*. The flow in the central region on the disk is *laminar*, while that in the outer peripheral region is *turbulent*. Although the oil film pattern corresponds reasonably well to the flow direction close to the surface, i.e. the local direction of wall shear stress, it indicates an exaggerated radial velocity component due to the centrifugal force acting on the oil film itself (Fig. 33).

Fig. 28. Taylor vortices with a low rotational speed of the inner cylinder.

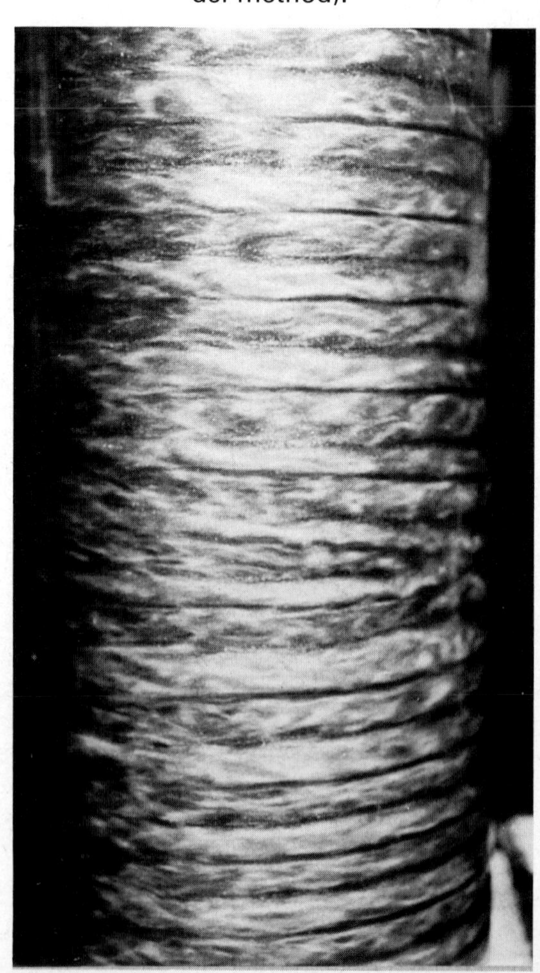

Fig. 29. Taylor vortices with a high rotational speed of the inner cylinder (water, speed of rotation of the inner cylinder: 2 rpm in Fig. 28, 30 rpm in Fig. 29, the outer cylinder at rest, OD of the inner cylinder 126 mm, ID of the outer cylinder 146 mm, aluminium powder method).

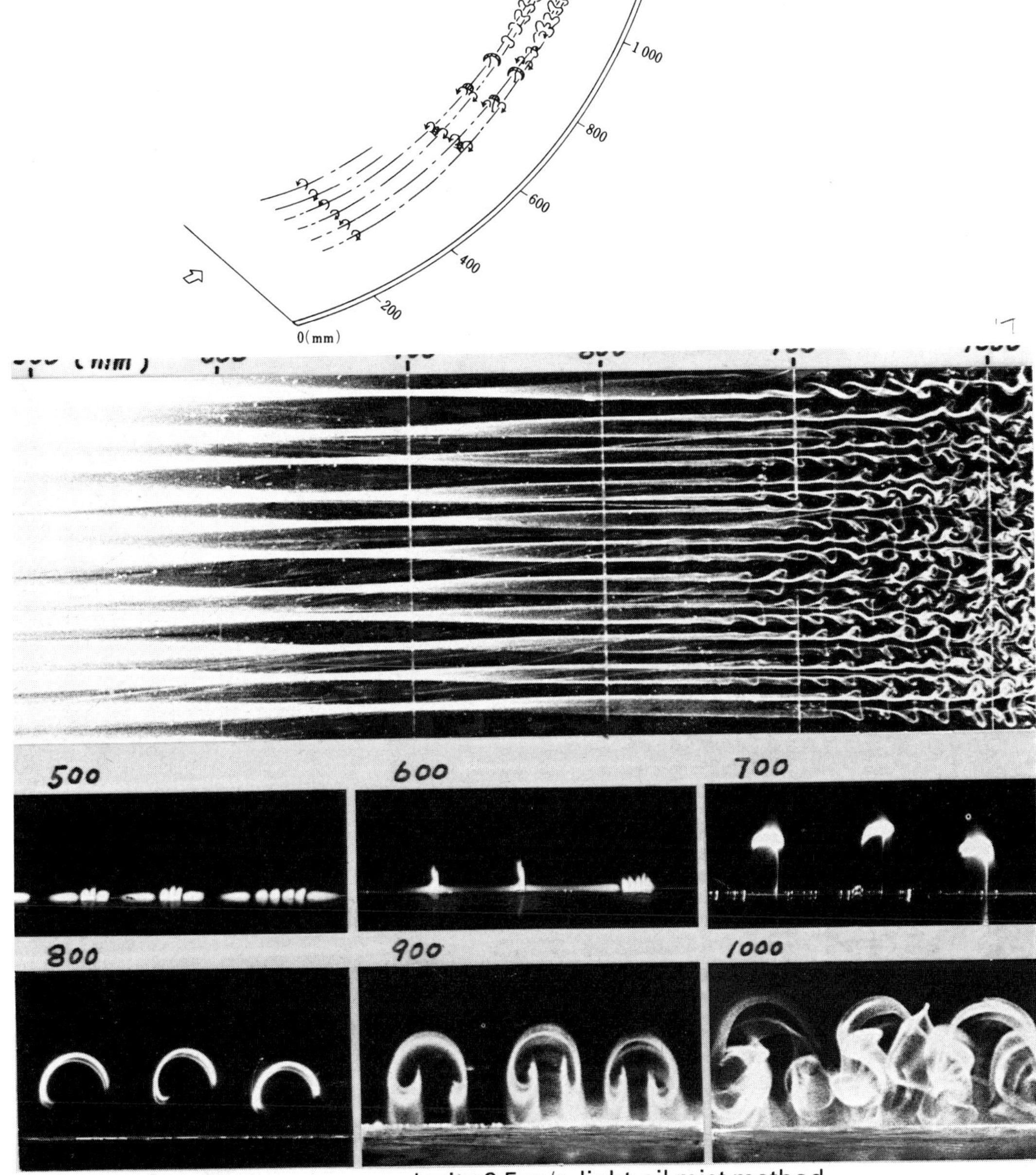

Figs. 30 and 31. Air, free stream velocity 2.5 m/s, light-oil mist method.

Fig. 32. Water, upper: velocity 11 cm/s, $Re = 1.5 \times 10^3$, middle: velocity 17 cm/s, $Re = 2.34 \times 10^3$, lower: velocity 54 cm/s, $Re = 7.5 \times 10^3$, pipe ID 14 mm, dye injection method.

The structure of a turbulent boundary layer

Downstream of the laminar to turbulent transition, the turbulent boundary layer along a flat plate continuously grows and eventually reaches a full-developed state. Figure 34 is an instantaneous flow visualization of the turbulent boundary obtained by the *hydrogen bubble method* with a fine electrode wire fixed normal to the wall, from which hydrogen bubbles are generated periodically. The horizontal flat plate is located at the bottom of the photograph. The instantaneous shape of the turbulent fluid suggests that the boundary layer consists of large vortical *bulges* of the scale of boundary layer thickness δ rolling downstream on the wall. The outer edge of the boundary layer, which clearly divides the turbulent and non-turbulent flow regions, has a complex corrugated shape. The region close to the wall seems to be strongly influenced by the presence of the solid wall. The fluid decelerated near the wall is periodically lifting away towards the outer layers in an event of some violence, while high-speed fluid rushes in towards the wall to replace it. These coherent motions in the near-wall region are known as the *bursting process*, and appear to be a primary mechanism for the production of *turbulence* and *turbulent shear stress*.

In Fig. 35 the structure of a turbulent boundary layer at various distances from the wall is visualized by the hydrogen bubble method. Here, a wire electrode is parallel to the wall and perpendicular to the flow direction. The location of the wire is shown in an attached figure of the *wall-plot* of the velocity profile. The velocity profile in the inner region of the boundary layer is well represented by the *law of the wall*, i.e. $u^+ = f(y^+)$ where $u^+ = u/u_\tau$, $y^+ = u_\tau y/\nu$ and $u_\tau = (\tau_w/\rho)^{1/2}$, *friction velocity*, respectively.

In Fig. 35, photographs A and B are of the *viscous sublayer*, while photographs C and D are of the *buffer layer* and the *logarithmic region*, respectively. The instantaneous velocity distribution in the spanwise direction is more or less periodic in the viscous sublayer and the so-called *low-speed streaks*, elongated in the streamwise direction, appear there. In the buffer layer violent mixing of fluid can be observed, and in the region further away from the wall the linear scale of the turbulence increases and the streaky structure is entirely lost.

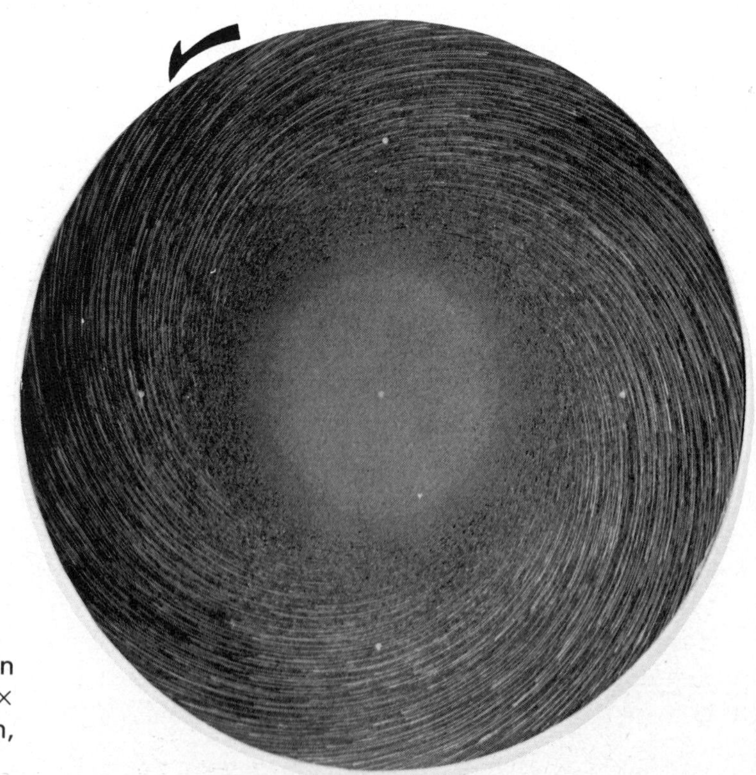

Fig. 33. Water, rotation speed 800 rpm, $Re = 1.9 \times 10^6$, disk diameter 300 mm, oil film method.

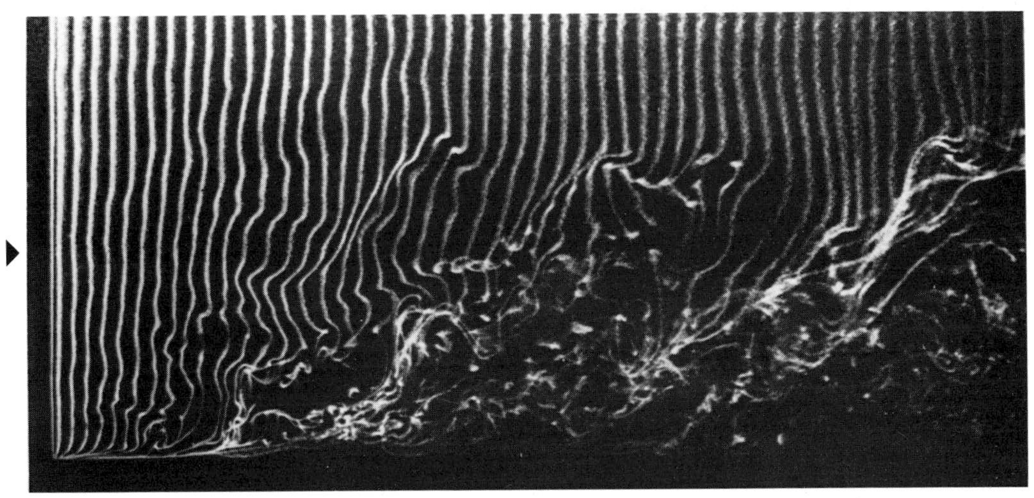

Fig. 34. Turbulent boundary layer along a flat plate
(water, free stream velocity 20.4 cm/s, $Re_\theta = 990$, hydrogen bubble method).

Velocity profile in the boundary layer (semi-log plot)

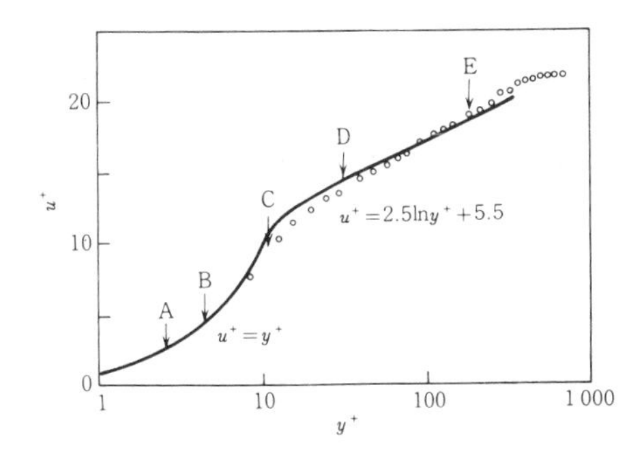

Fig. 35. Turbulent boundary layer structure
(water, free stream velocity 20.4 cm/s, $Re_\theta = 990$, hydrogen bubble method).

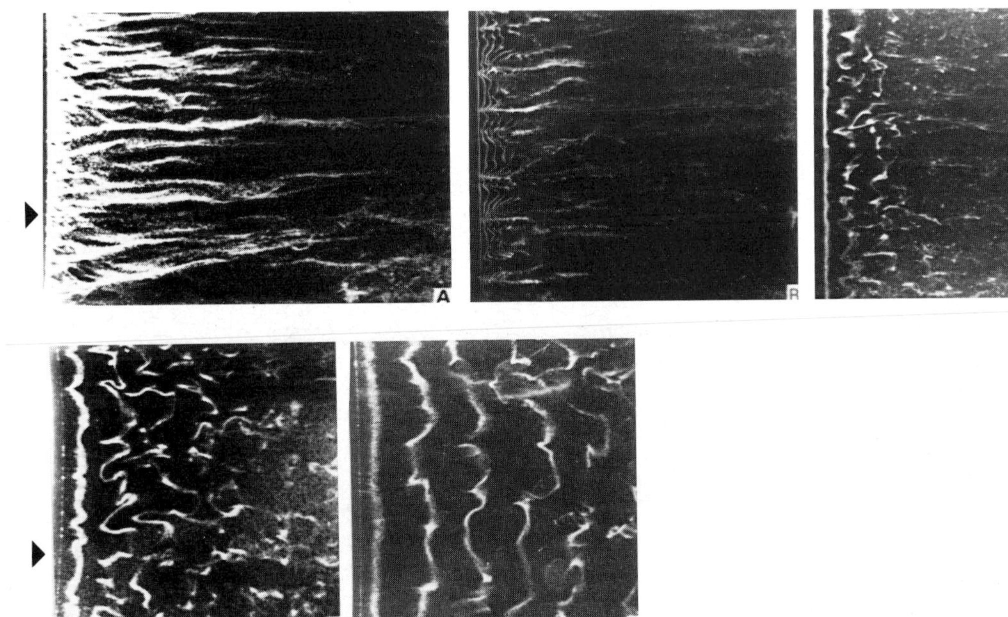

Laws of Similarity

Reynolds' law of similarity

In an incompressible fluid flow, disregarding gravity forces, the flow is governed by inertia forces and viscous forces. Flows around geometrically similar bodies are, in such a case, dynamically similar if the ratio of inertia force/friction force is same for both flows. A measure of this ratio is given by the *Reynolds number*, *Re*:

$$Re = \frac{\text{inertia force}}{\text{friction force}} = \frac{UL}{\nu}$$

where U is a characteristic velocity of the flow, L is a characteristic dimension of the body and ν is the kinematic viscosity of the fluid. In planning a model experiment, or in using any existing experimental data, therefore, the Reynolds number of the tests must be as near as possible to that of actual device or situation. If this is impracticable the effect on the flow of the difference of Reynolds number must be carefully examined.

The flow over the upper surface of the main wing of an aeroplane model is shown in Fig. 36, the flow being made visible by means of surface tufts. Shown in Fig. 37 is the flow over the outer wing of the full sized aircraft at roughly the same angle of attack (the region concerned is marked in Fig. 36). This picture was taken during flight test, through a window indicated in Fig. 36 by a triangle. In spite of the difference in Reynolds number, the two patterns of surface flow in this case are very similar.

Froude's law of similarity

On a free surface, gravitational waves caused by the motion of a ship result in wave-making resistance. The development of the waves is governed mainly by inertia forces and gravity. The ratio of the inertia forces to the gravity forces, given by $U/\sqrt{gL}$ where U is velocity of flow, L the length of the ship and g the acceleration due to gravity, is called the *Froude Number*, *Fr*. The gravity wave pattern, surrounding two geometrically similar objects moving on a free surface, is similar if the Froude numbers are the same. This is called Froude's law of similarity.

When the Froude number is the same, and hence the wave pattern is the same in two cases, then the wave-making resistances, expressed in appropriate dimensionless form, are also the same.

The photograph in Fig. 38, was taken from a helicopter, and shows the pattern of waves formed by a passenger ship sailing in the Inland Sea of Japan. Figure 39 shows the pattern of waves produced by a model ship in a test tank running at the same Froude number as the passenger ship. Visualization is performed by aluminium powder dispersing over the water in the tank. Although the pattern of foam near the passenger ship obscures the wave pattern, the waves outside the region of the foam can be seen to be similar in the two cases.

Figures 40, 41 show wave shapes along the sides of two similar model ships towed at the same Froude number in two different water test tanks. Because of the differences between the test tanks, towing arrangements and lighting and viewing angles, the diverging surface waves are not clearly shown, but the wave shapes on the sides of the model ships can be seen to be similar.

Fig. 36. Flow over the upper surface of the main wing
– – – – – Wind-tunnel test
(air, $U = 40$ m/s, chord length = 195.6 mm, attack angle = 10°, $(Re)_{\text{nominal}} = 0.5 \times 10^6$, surface tuft method).

Fig. 37. Flow around the outer wing on the starboard side
– – – – – Flight test
(air, $U = 43.7$ m/s, chord length = 1505 mm, attack angle = 12°, $Re = 5\times 10^6$, surface tuft method).

Fig. 38. Pattern of waves formed by a passenger ship (sea water, sailing speed 9 m/s, ship length 80 m).

Fig. 39. Pattern of waves produced by a model ship (water, running speed 1.15 m/s, model length 1.30 m, aluminium powder dispersing method).

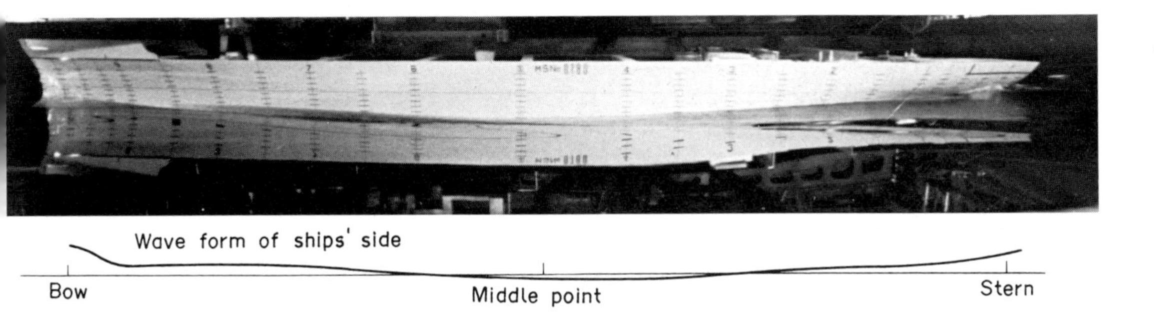

Fig. 40. Pattern of waves produced by a large size model ship (water, running speed 2.99 m/s, model length 10.0 m).

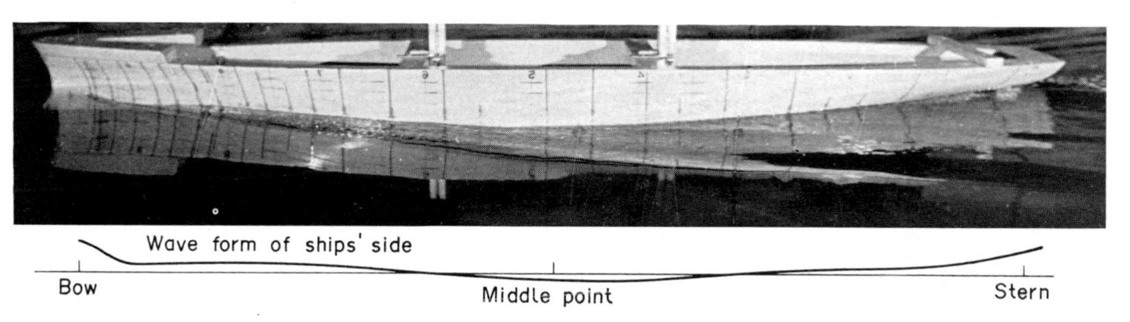

Fig. 41. Pattern of waves produced by a small size model ship (water, running speed 1.35 m/s, model length 2.0 m).

Compressible Flow

Subsonic and supersonic flows

In subsonic flow, changes of flow properties which arise on the flow boundary or in the flow itself, propagate as disturbances over the whole flow field, whereas in a supersonic flow only part of the flow field is modified. A disturbance of infinitesmal amplitude propagates as a spherical sonic wave with the sonic speed a relative to the fluid moving with velocity U. The ratio U/a is called the *Mach number* M and its value indicates whether or not compressibility effects will be of primary importance (to the flow behaviours).

In Figs. 42, 43, 44 and 45, schlieren photographs show compressible flows over a wedge, illustrated in Fig. 42(a). The photographs also show the supports for the wedge, and they appear as two nodules. They do not influence the flow but obstruct the view a little.

In *subsonic flows*, as shown in Fig. 42, the flow velocity is lower than the sonic speed i.e. the sonic waves emitted by a source P, shown on the inset of Fig. 42, eventually extend to infinity in every direction, though the wave front array is distorted in the downstream direction as indicated on this inset diagram. In the picture, waves originating from the vortex shedding activity at the trailing edge of the wedge, can be seen propagating upstream both over and under the wedge. In the waves propagating downstream, the density gradients are too small to produce a visible image in the schlieren system. The upstream propagating waves in this case excite a boundary layer instability which exerts a controlling influence on the vortex shedding process.

In *supersonic flows*, shown in Fig. 43, the flow velocity is higher than the sonic speed, i.e. $M > 1$, and the disturbances are propagated only downstream. Hence the influence of a disturbance is restricted within a cone which has its vertex at the source P, shown on the inset of Fig. 45, enveloping the spherical waves. The source and the flow upstream of the cone are not influenced by the disturbance. In two-dimensional flow, the cone degenerates to a pair of two *Mach lines* each crossing the streamline with the Mach angle $+\mu$ and $-\mu$. Numerous Mach lines are visible issuing from the wedge surface. If the flow is compressed along the stream line, the Mach lines converge on each other forming an *oblique shock wave*, and, if the flow is expanded, the Mach lines diverge constituting a *Prandtl–Meyer wave*, as shown in the insets. It is visible in the figure that both waves issue from the apex and the shoulder of the wedge respectively.

In *transonic flows*, as shown in Fig. 44, the velocity is approximately equal to the sonic speed i.e. $M \simeq 1.0$. The flow field in Fig. 44 is composed of subsonic flow behind the bow shock wave and low supersonic flow restored by the flow through the expansion from around the wedge.

The schlieren picture shown in Fig. 45 is quite different from the three preceding ones. Figure 45 is obtained using a shock tube where an initially stationary gas is suddenly accelerated to approximately uniform flow by a moving shock wave travelling into the gas. This is a high strength shock wave and can be seen as the vertical line across the picture through the right-hand half of the wedge. When the moving shock wave interacts with the model wedge reflections and interactions take place, some of which are quite complex. However, the v-shaped oblique shock waves in the vicinity of the leading edge of the wedge may be

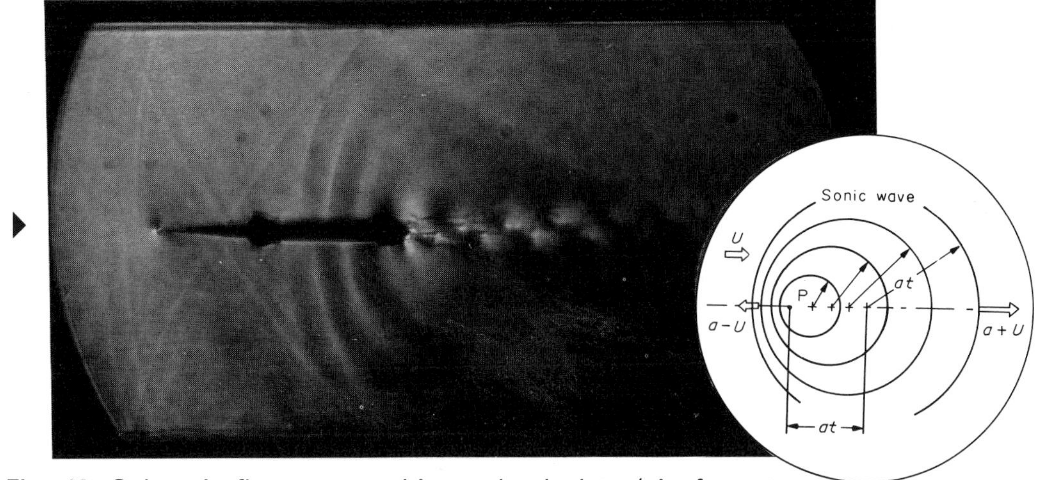

Fig. 42. Subsonic flow past a thin wedged plate (air, free stream velocity 216 m/s, $Re = 5.4 \times 10^5$, $M = 0.5$ chord length 59 mm, colour-schlieren method).

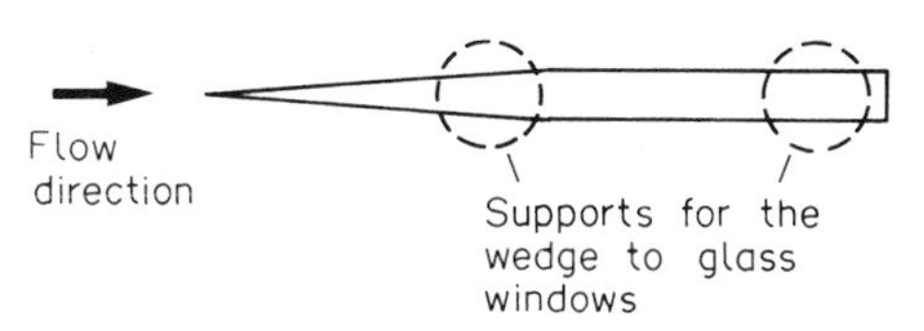

Fig. 42(a). Illustration of wedge.

Fig. 43. Supersonic flow past a thin wedged plate (Freon R-114, free stream velocity 189 m/s, $Re = 3.8 \times 10^6$, $M = 1.4$, chord length 59 mm, colour-schlieren method).

compared with the corresponding region in Fig. 43. Particularly clear in Fig. 45 are the Mach lines, visible on the left hand side of the picture, and originating from small irregularities on the upper and lower walls of the shock tube. The irregularities in this case are due to liquefaction of the vapour flow on the cold walls.

Interaction of shock waves and boundary layers in duct flow

A reduction of supersonic flow to subsonic flow in a duct usually takes place across shock waves, which interact with the boundary layers along the duct wall. The interaction becomes significant as the shock wave

Fig. 44. Transonic flow past a thin wedged plate (air, free stream velocity 219 m/s, $Re = 6.0 \times 10^5$, $M \approx 1.0$, chord length 59 mm, colour-schlieren method).

Fig. 45. Formation of Mach line, or weak shock wave, by weak disturbance is supersonic flow (freon R-114, free stream velocity 190 m/s, $Re = 3.4 \times 10^6$, $M \approx 1.4$, chord length 59 mm, colour schlieren method).

becomes stronger and as the boundary layer grows thicker. The shock wave then no longer remains a thin and a plane wave. Due to the separation of the boundary layer, the wave is bifurcated near the wall and a train of shock waves is induced. The wave train is called a *pseudo shock wave*, since it functions as a whole almost as the equivalent to a single normal shock wave.

Figures 46 and 47 show the interactions arising in a straight duct of constant cross-sectional area. The shock wave train is observed downstream of the leading wave. In the case of a weak interaction, the normal configuration of the leading wave is still maintained in the central region, despite the bifurcation in the wall region. Such a pseudo wave is called *λ-type*. In a strong interaction, the triple point of the bifurcation reaches the tube axis, and the normal part does not exist in the leading wave. Such a wave is called *x-type*.

Supersonic flow around a jet engine inlet

The frontispiece picture ix-top shows the shock wave pattern appearing in supersonic flow into and around a jet engine inlet. At a high throttle ratio, the shock wave is stable as seen in the picture. However, at a throttle ratio less than a critical value, a λ-type shock wave appears oscillating significantly ahead

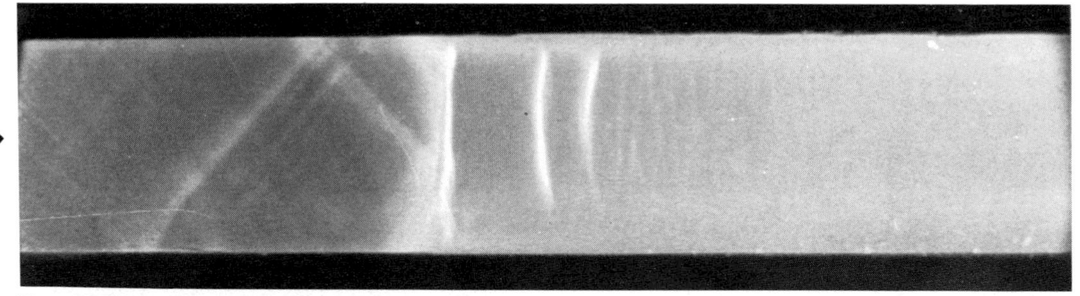

Fig. 46. Shock wave and boundary layer interaction duct flow.

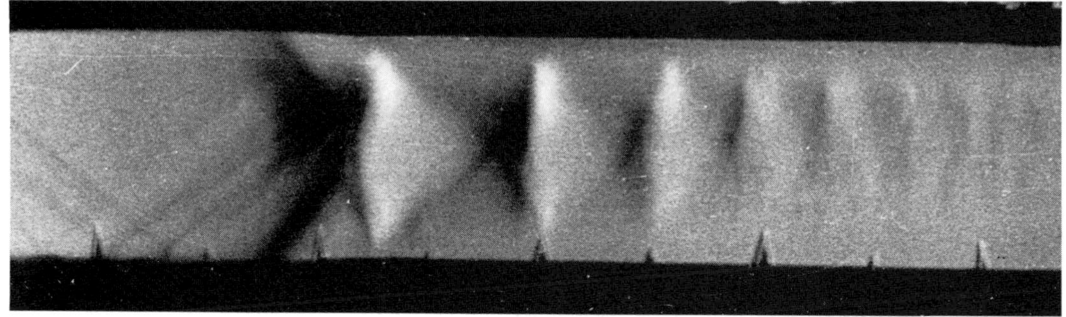

Fig. 47. Shock wave and boundary layer interaction in duct flow – strong interaction (air, flow velocity ahead of the leading wave 390 m/s [46], 490 m/s [47], duct cross-section 60 × 60 mm, schlieren method).

Fig. 48. Supercritical flow and hydraulic jump of shallow water from spillway (water, supercritical flow velocity 50 cm/s).

of the intake. The oscillation causes troublesome variations of pressure in the flow to the engine and is called *inlet buzz*.

Hydraulic jump

Observations of *shallow water waves* may be helpful in obtaining a qualitative understanding of gas dynamic flows. Figure 48 shows a *hydraulic jump*, which is an analog of a normal shock wave. The flow leaving the steep spillway is accelerated to supercritical in a sense that the velocity exceeds the gravity wave speed. By crossing the jump, the flow becomes slow and deep, and a uniform subcritical flow is attained in which

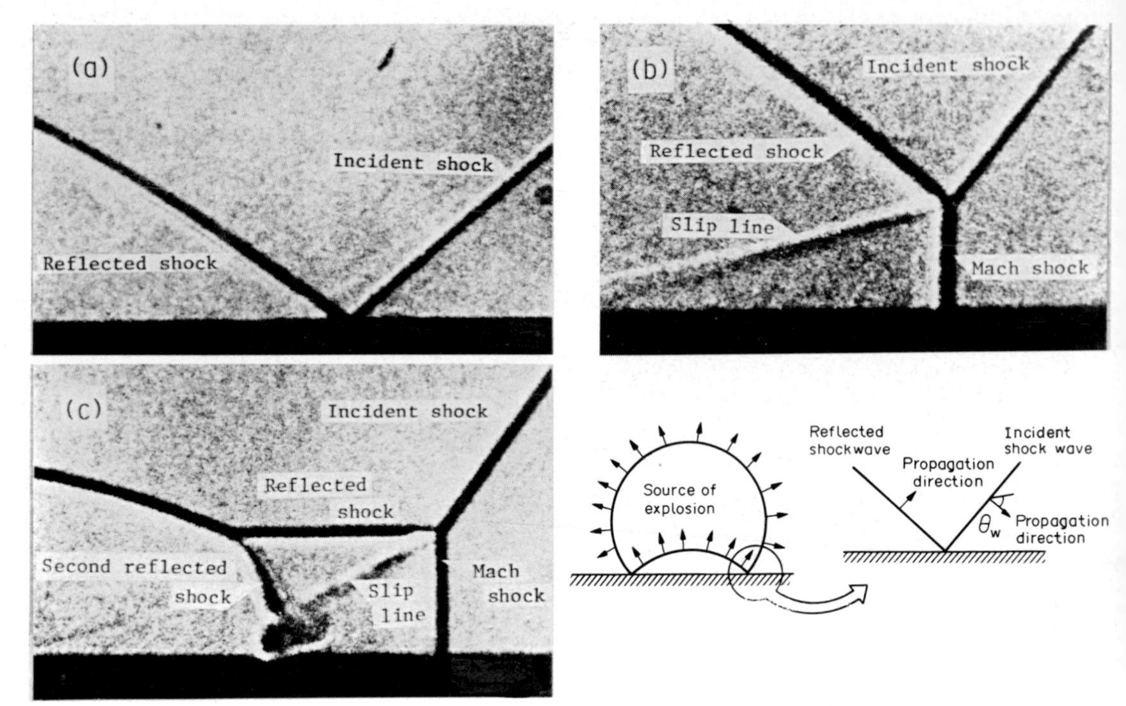

Fig. 49. Plane shock wave impinging on an inclined wall (carbon dioxide gas, shadowgraph method).

the velocity is less than the wave speed. The flow change produced by the jump is dissipative and, as in the gas dynamic analogy, the equivalent of a rarefaction shock is not possible.

Reflection of a plane shock wave from an inclined wall

In the instantaneous shadowgraphs Fig. 49, the shock wave propagating into a stationary gas in the direction normal to the wave front impinges obliquely on a wall. The reflection takes a configuration of *regular reflection*, *single Mach reflection*, or *double Mach reflection*, depending on the Mach number *Ms* of the impinging wave, the wall inclinaticn θ_w to the direction of the shock propagation, and thermodynamic properties of the gas.

The regular reflection (a) is usually observed when θ_w is large. In the case of the single Mach reflection (b), the impinging wave cannot reach the wall, and a *Mach shock wave* stands perpendicularly to the wall. A slip surface issues from the triple point where the incident, the reflected and the Mach waves intersect. This configuration occurs in cases where both *Ms* and θ_w are low. Finally, the double Mach reflection (c) appears at a considerably higher value of *Ms* and with values of θ_w within a particular range. The reflected wave is kinked at a certain distance from the triple point, and a second reflection wave is formed. The slip surface curls up in this case.

Supersonic flow through a Laval nozzle

The supersonic flow pattern through a *Laval nozzle*, or a convergent-divergent nozzle depends on the ratio ϕ of the inlet stagnation pressure to the nozzle back pressure. As ϕ is increased above 1, the critical condition is first reached at the nozzle throat where the flow velocity first becomes equal to the local sonic speed. A normal shock wave then forms at a section slightly downstream of the throat. This shock wave then moves toward the nozzle exit forming the pseudo shock wave due to the boundary layer interaction as shown in Fig. 50. With further increase in ϕ the shock wave, moves outside the exit, as shown in Fig. 51, taking a form of an oblique shock pair with a symmetric inclination to the stream line (the shock originating in the upper lip is white in this photograph, making it less obvious than its pair from the lower lip). The back pressure is higher than the pressure of the flow at exit,

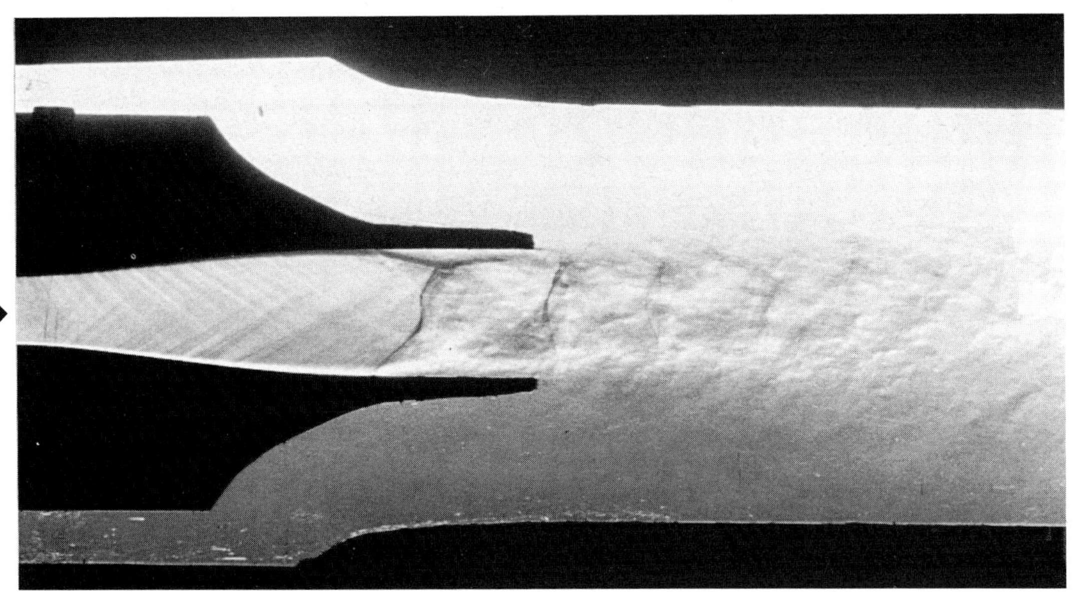

Fig. 50. Shock-in-nozzle flow through a Laval nozzle
[air, stagnation pressure of the incoming flow 210 kPa, flow velocity ahead of shock wave 490 m/s (Mach number 1.9), throat cross-section $5.0 \times 32.5\ mm^2$, exit cross-section $30 \times 32.5\ mm^2$, schlieren method].

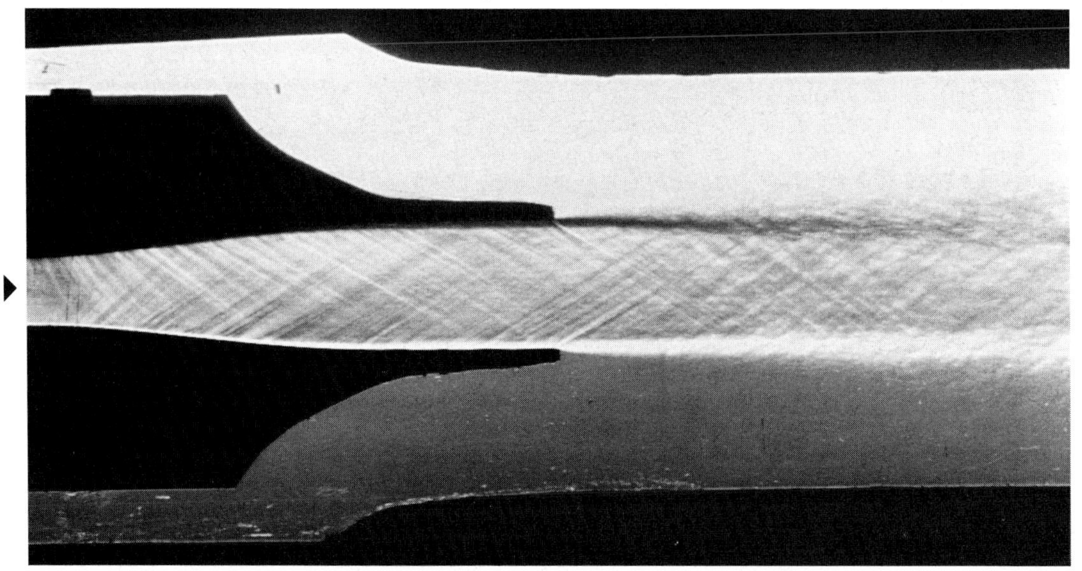

Fig. 51. Over-expanding flow through a Laval nozzle
[air, stagnation pressure of the incoming flow 294 kPa, nozzle back pressure 72 kPa, exit flow static pressure 38 kPa, exit flow velocity 510 m/s (Mach number 2.0)].

and is matched by the flow after the shock pair has been passed: thus the flow is called an *over-expanded flow* . As ϕ is increased still further, the oblique wave is weakened until it degenerates to a Mach wave, and the *correctly expanding flow* of Fig. 52 is attained, where the back pressure is reached directly by the flow at exit. With a further increase in ϕ, the flow emerges with an excess pressure over the surrounding fluid; i.e. *under-expanded flow*. Outside the nozzle exit, the jet flow further expands and diverges as can be seen in Fig. 53 crossing an expansion fan pair which issues from the corners of the nozzle exit.

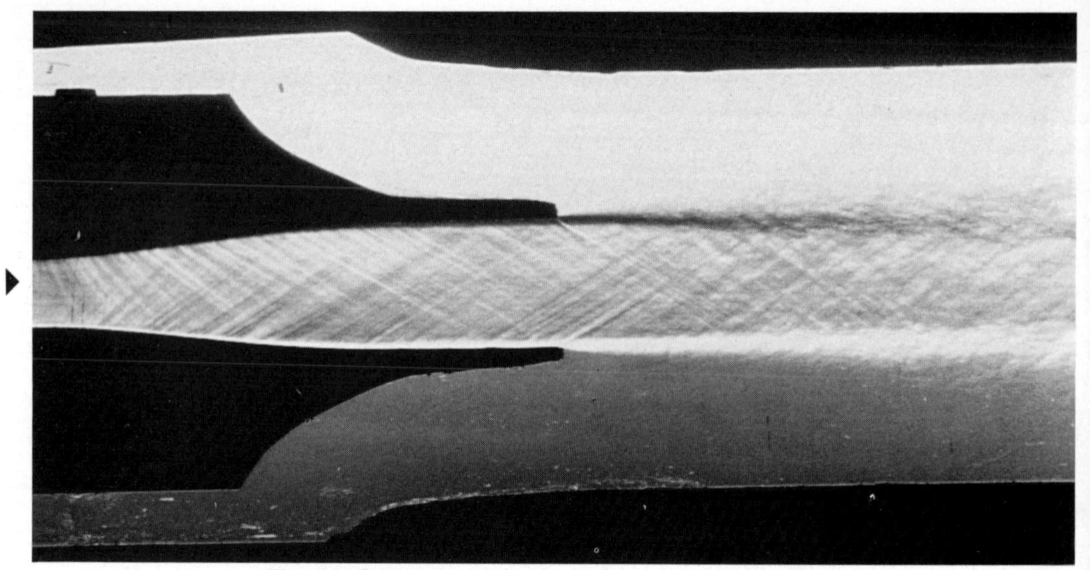

Fig. 52. Correct-expansion of a Laval nozzle flow
[air, stagnation pressure of the incoming flow 364 kPa, nozzle back pressure 47 kPa, exit flow static pressure 47 kPa, exit flow velocity 510 m/s (Mach number 2.0)].

Fig. 53. Under-expansion of flow by a Laval nozzle
[air, stagnation pressure of the incoming flow 395 kPa, nozzle back pressure 40 kPa, exit flow static pressure 51 kPa, exit flow velocity 510 m/s (Mach number 2.0)].

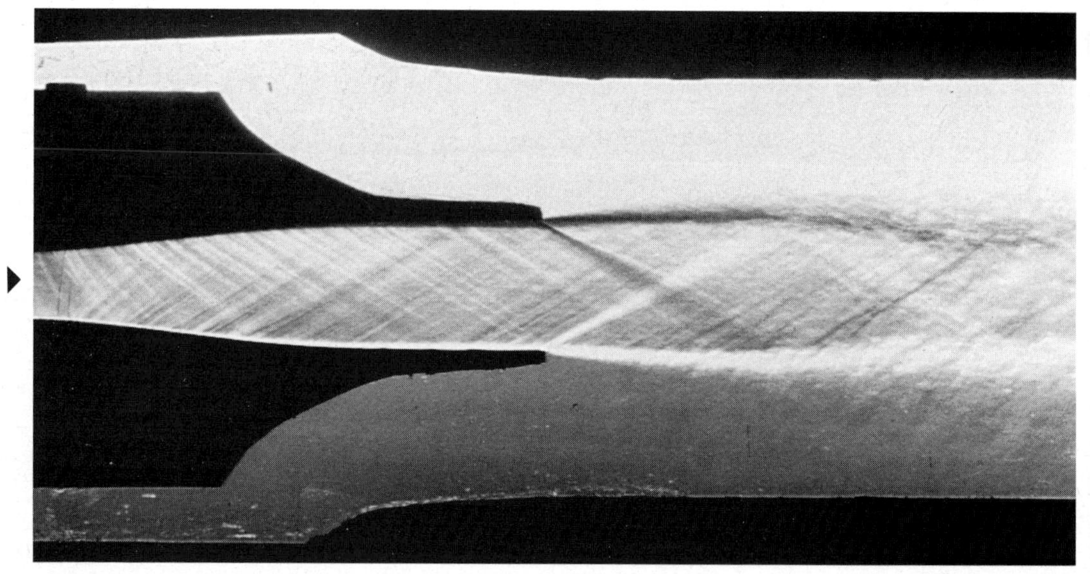

Transonic flow around an aerofoil

In transonic flow, the shock waves formed on an aerofoil surface induce boundary layer separation, causing a significant increase in aerodynamic drag. Figures 54 to 59 show the flow configurations on an aerofoil which is designed to suppress shock formation up to a high Mach number.

In Fig. 54, where the free stream Mach number M is 0.706, the flow is everywhere subsonic and the boundary layer separates in a normal manner. At $M = 0.826$, the flow is slightly supersonic midway along the aerofoil, and a few weak shock waves and expansion fans have appeared. However, the separation is still moderate and the drag force remains

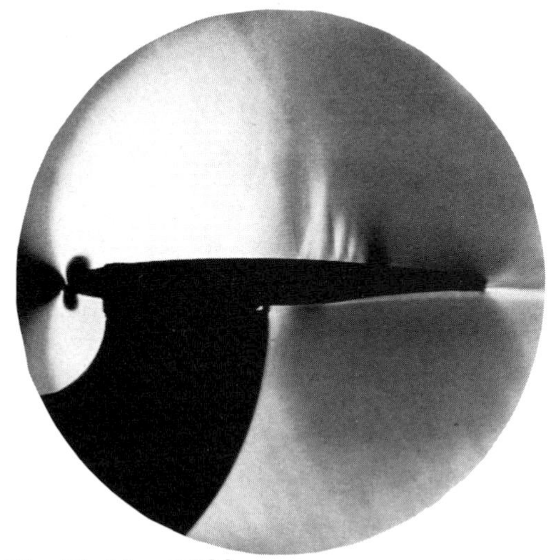

Fig. 54. $M = 0.706$.

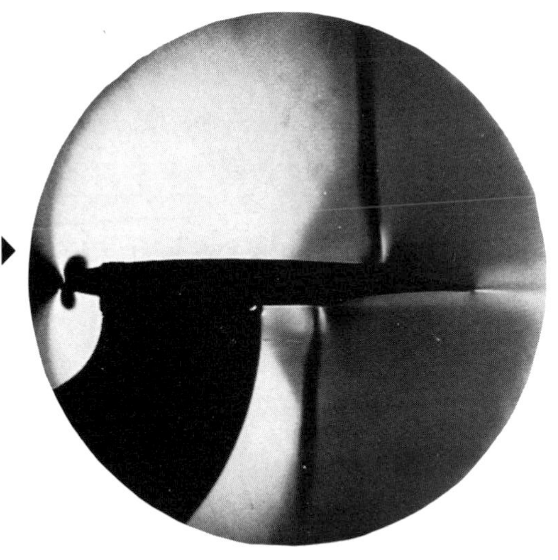

Fig. 55. $M = 0.826$.

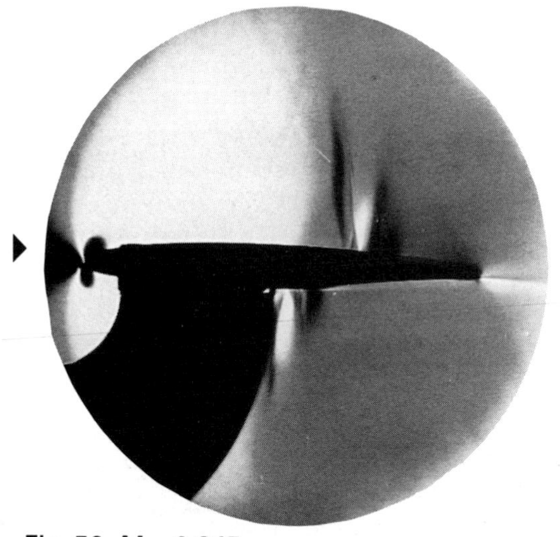

Fig. 56. $M = 0.845$.

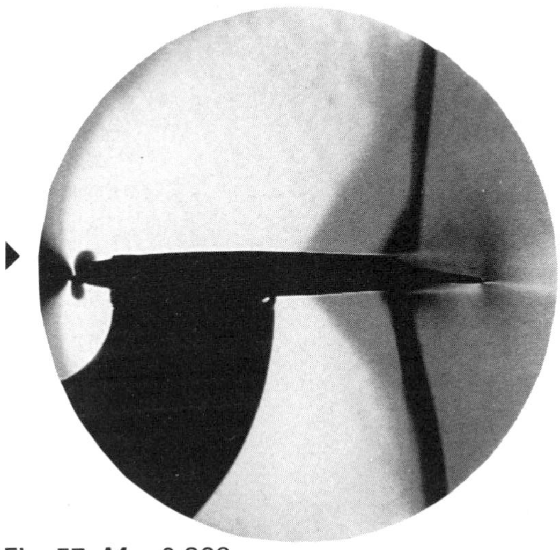

Fig. 57. $M = 0.866$.

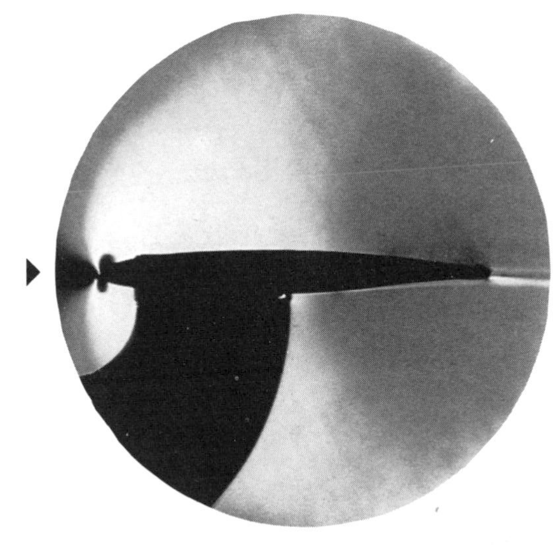

Fig, 58. $M = 0.930$.

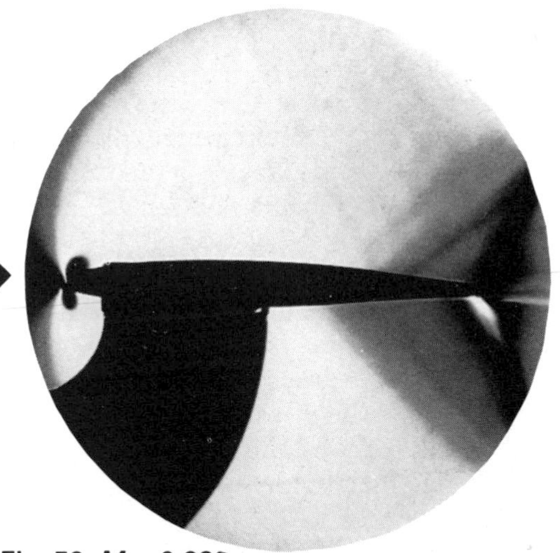

Fig. 59. $M = 0.992$.

(Air, attack angle 0°, schlieren method).

Fig. 60. Interferogram in constant-density Fringe Mode showing transonic flow through turbine cascade (air, exit velocity 445 m/s, chord length 33.6 mm, $Re = 5.2 \times 10^5$, Mach–Zehnder Interferometry).

at a low level. The flow at this stage is called the flow of the *drag-divergence Mach number*. In a range of $M = 0.845, 0.866$, Figs. 56 and 57, the shock waves coalesce to form a single but a strong normal shock, and the onset of the separation moves ahead of the wave. The boundary layer does not reattach to the surface, and a rapid increase of drag starts at this stage.

At $M = 0.930$, Fig. 58, the shock wave has grown up to an intense λ-type wave standing in the vicinity of the trailing edge. Due to the heavy separation, the wake is jet like and the highest drag coefficient is attained. As M approaches 1, Fig. 59, however, the shock waves reach the trailing edge and are reduced to oblique waves. Since the separation is restricted close to the trailing edge, the wake is thin and the drag begins to decrease. Finally, a bow shock wave is formed ahead of the leading edge as M exceeds 1, but this is not shown in the photographs.

Transonic flow through a turbine cascade

The three pictures, obtained by three different methods of visualization, show typical *transonic flow through a turbine cascade* with subsonic entry flow and supersonic exit flow.

An interferogram of the flow is shown in Fig. 60. The black and white fringes indicate lines of constant density (isochoric lines) within the flow. The flow direction is from lower left to upper right of the picture, and the exit velocity is about 5 times the inlet velocity, with an exit Mach number of 1.3.

The convex and concave surfaces of the turbine blades act as suction and pressure surfaces respectively. The resulting aerodynamic force is exerted in the upward direction of the picture plane, doing work on the cascade to generate rotational power.

Various aspects of the flow may be read from the interferogram: e.g. the gradual decrease of density from inlet to outlet, and the formation of a steep density gradient or shock wave where the constant-density lines fold together. The same flow is again shown in Fig. 61: This picture is obtained by the schlieren method, which reveals density *gradients* and is useful in giving an easily

Fig. 61. Schlieren picture of turbine cascade flow
(air, exit velocity 441 m/s, chord length 33.6 mm, $Re = 5.2 \times 10^5$ schlieren method).

interpretable picture of the flow involving waves of large density change. Shock waves develop from the trailing edge into the streams on both sides of the blades and the lower branch is reflected from the upper surface of the adjacent blade. The black and white zone issuing from the trailing edge, parallel to the stream, indicates the wake of the blade where the flow is retarded.

The surface wave pattern in a *shallow water flow* is shown in Fig. 62 as an analog of the compressible flow patterns. The cascade geometry is similar to that in the previous two pictures. The wave pattern downstream of the blades is very similar to the shock wave patterns in Figs. 60 and 61, except that the lower branch of the trailing edge wave does not impinge on the adjacent blade: this is because the equivalent of the Mach number, i.e. the *Froude number*, is higher than the air flow Mach number.

Trailing shock wave and boundary layer separation in a transonic turbine cascade

The flow expands from subsonic to supersonic along the passage between the blades, and shock waves originate from the trailing edge of each blade. Interaction of these waves with the boundary layer on the neighbouring blade is a major cause of cascade performance failure. Flow patterns for exit flow Mach numbers M between 1.2 and 1.6 are shown in Fig. 63.

At $M = 1.2$ and 1.3, the shock wave STP issuing from each blade impinges on the suction surface of the adjacent blade, causing a significant separation of the boundary layer sb. The reflected wave SR reflects back yet again from the wake of the original blade and interacts with the already separated boundary layer of the lower blade. Hence, the shock wave STS developing from the trailing edge toward the suction side of the blade becomes very strong. The pressure reduction on the suction side of the blade is reduced, and the thickened wake leads to a large loss of total pressure.

For M higher than 1.4, the wave STP leans away downstream, and the location of impingement on the lower blade moves out beyond the trailing edge. The boundary layer is no longer disturbed, and the increase of pressure-loss with increasing Mach number in the cascade is reduced.

Fig. 62. The turbine cascade flow observed in shallow water table (water free stream velocity 45 m/s, chord length 100 mm, Froude number 1.7, $Re = 4.5 \times 10^4$).

Supercritical flow through a conical plug valve

Several different configurations of high speed jet are found in the throttling of high pressure gas by valves. At pressure ratios, P_o/P_a, higher than the critical ratio, the jets are supersonic and produce significant noise and vibration. Typical patterns of an annular *flow throttled by a conical plug valve* are shown in Fig. 64.

Under high supercritical conditions, an expansion wave E issues from the throat, as shown in the diagram, and a compression wave C is formed by its reflection at the free surface. Then the boundary layer on the plug is separated by the compression, and a shock wave Ss is generated. At $P_o/P_a = 7$, the flow reattaches on the plug surface. However, at $P_o/P_a = 10$, the flow passes over the plug without reattachment. At this stage, the turbulent mixing along the free shear zone, and hence also the aerodynamic noise, both reach their most intense levels. The thrust on the plug fluctuates heavily due to the instability of the fully separated wall jet.

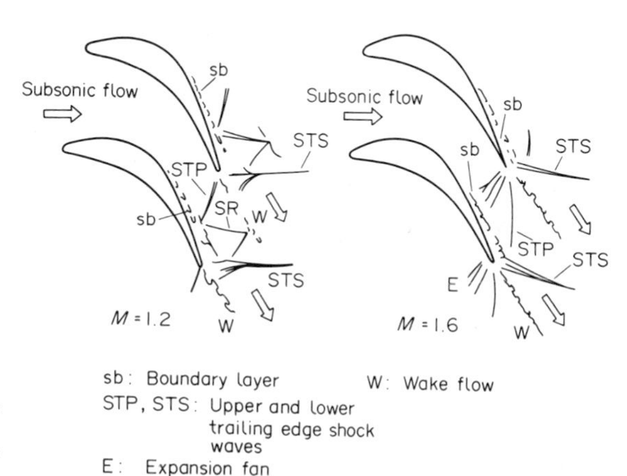

Fig. 63. Trailing edge shock waves in turbine cascade flow (superheated steam, exit flow Mach number 1.2–1.6, $Re = 3.5 \times 10^5$, chord length 52 mm, colour-schlieren method).

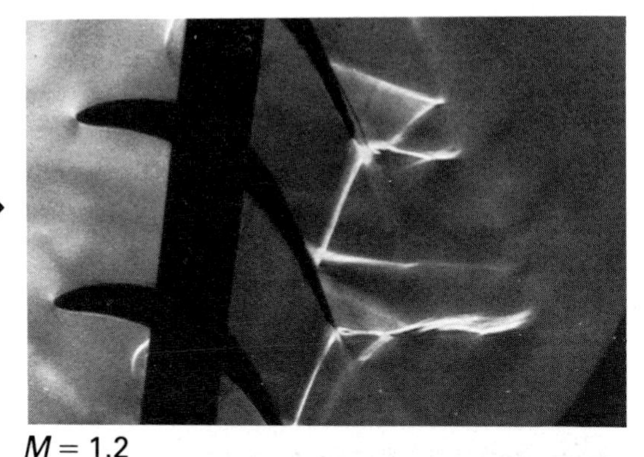

$M = 1.2$

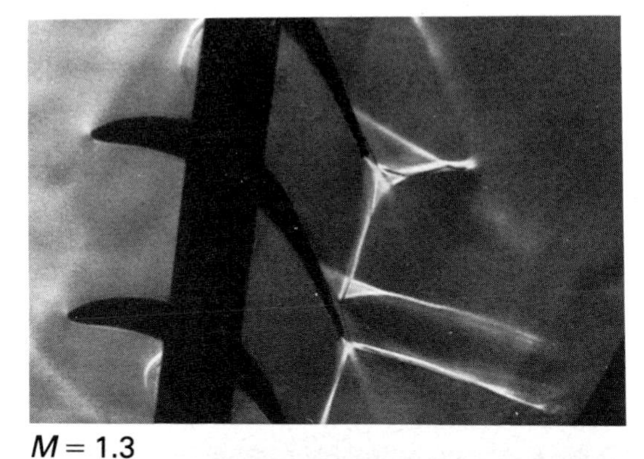

$M = 1.3$

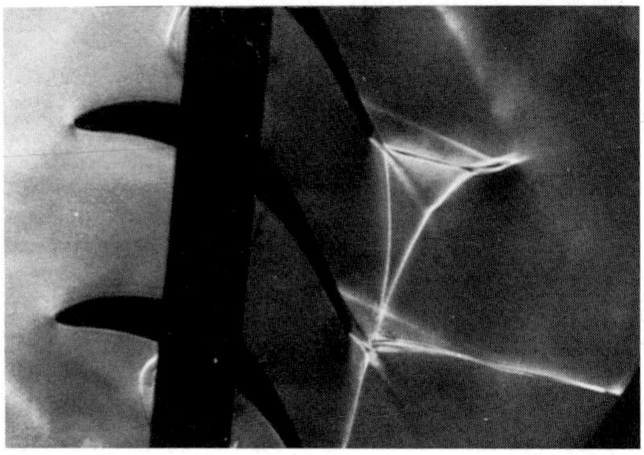

$M = 1.4$

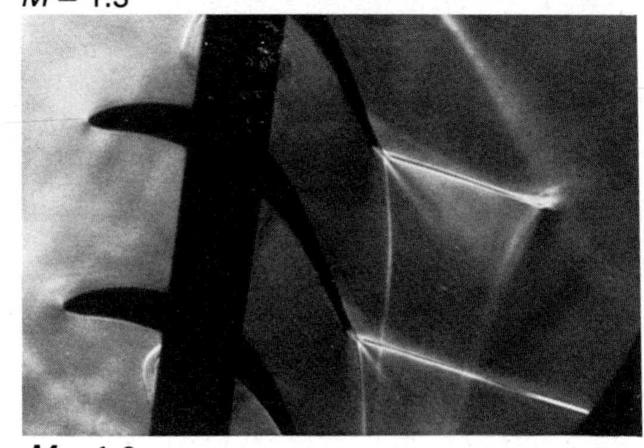

$M = 1.6$

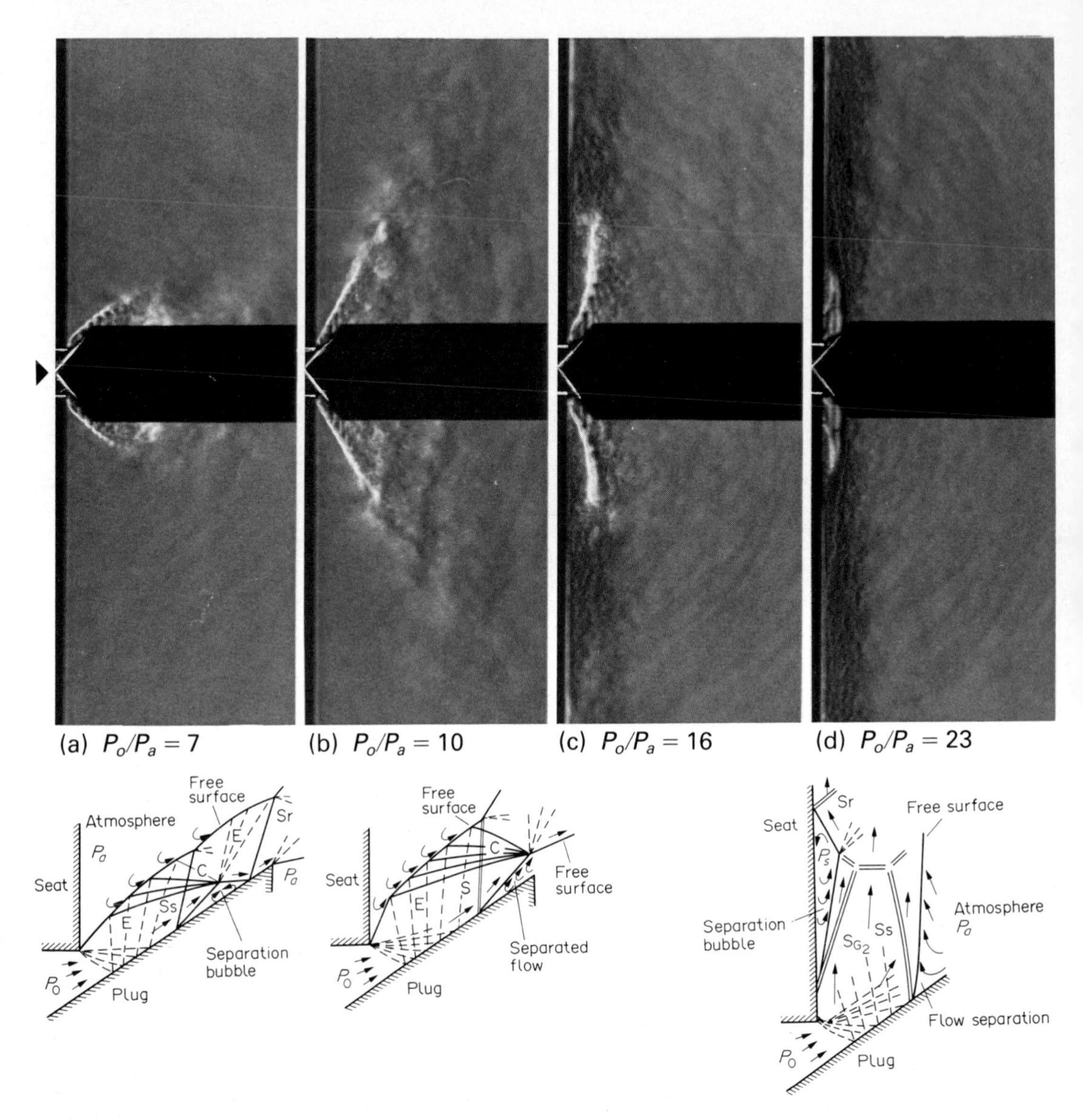

Fig. 64. Supercritical jet flow issuing from a conical plug valve [dried air, diameter of the valve seat 10 mm, opening area ratio 0.4, vertex angle of the plug 100 deg., plug diameter 20 mm, $Re = (1–4) \times 10^5$, colour-schlieren method].

As P_o/P_a is increased further, the annular jet suddenly attaches on the seat plane as shown in Fig. 64(d). This switching is possibly triggered by the entrainment into the separated jet, which is inherently unstable, of the air between the jet and the seat. The noise is much reduced by the transition, since the free surface of this pattern originates from a separation with a less intense velocity shear. The pattern (c) appears in the intermediate stages, before the pattern (b) is restored as the pressure ratio is reduced.

Jets

Mixing layer between two flows

When two parallel streams, which move at different speeds in the same direction, meet, a *mixing layer* develops which accommodates the velocity difference, spreading laterally in the downstream direction.

Figure 65 shows the case of a two-dimensional confined *jet*, of velocity $U_1 = 2.0$ cm/s discharging from a nozzle into a surrounding fluid of velocity $U_2 = 0.25$ cm/s. The velocity profiles of these two flows are visualized by the *hydrogen-bubble method*, showing the details of their confluence in the downstream direction.

Figures 66 and 67 show the case of $U_2 = 0$. The jet has a uniform velocity at the nozzle exit, but, as it flows downstream, it is mixed with the surrounding fluid at rest, decreasing gradually the potential core of uniform velocity, eventually developing velocity profiles similar to those of Fig. 65.

Collision of two opposed jets

Figure 68 is a picture visualized by the *hydrogen-bubble method*, showing how two opposed circular jets collide with each other. Very thin radial flow is observed after *collision*, similar to that found when a *jet* impinges on a flat plate.

Axisymmetric underexpanded jet

A *supersonic jet* is discharged from a *convergent nozzle* when the stagnation pressure of the gas is higher than a critical value above the atmospheric pressure.

The pressure of the jet at the nozzle exit is still higher than the surrounding atmospheric pressure, so that expansion of the jet must occur just downstream of the exit. This phenomenon is called *underexpansion*.

In the underexpanded jet, the wave and flow patterns undergo cyclic changes, compression wave after expansion and vice versa, as shown in Fig. 69.

Coanda effect of a wall jet

A tangential jet discharged from a nozzle

Fig. 65. Mixing layer between two flows (0.01% solution of salt, nozzle diameter 15 mm. *Re* = 300, hydrogen-bubble method).

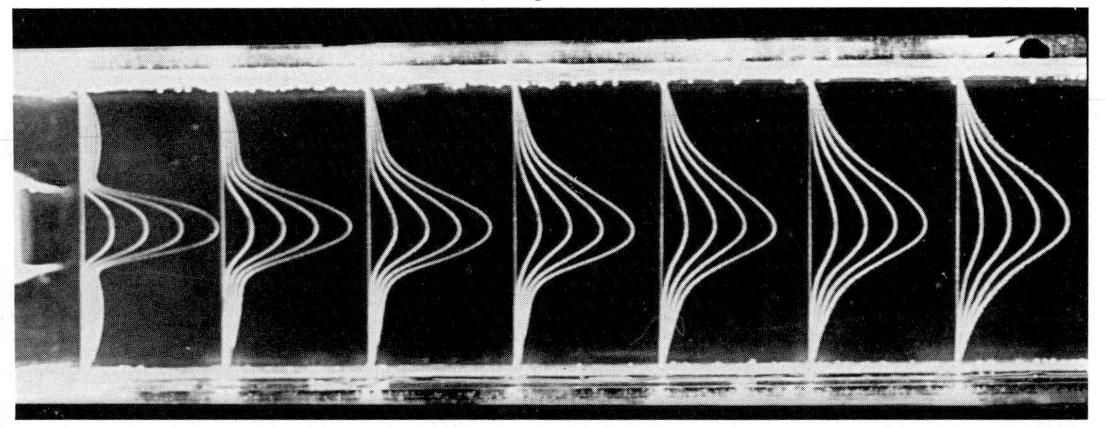

on the surface of a circular cylinder flows around the cylinder surface without separation until it detaches from the cylinder after being bent through an angle of 135°.

In Fig. 70 the flow is visualized by means of a white fog of *dry ice*, showing that part of the jet flow is carried round the cylinder to a point on the cylinder surface making an angle of about 210° with the nozzle exit.

This phenomenon is attributed to *Coanda effect*: that is, the turbulent boundary layer developed on the cylinder surface entrains the surrounding fluid to make the pressure at the cylinder surface negative, so that the jet is pulled round the cylinder by a transverse pressure gradient. When the characteristic Reynolds number, $Re^\star = U_j\sqrt{R}h/2/\nu$ is larger than (1.4–3.5) × 10^4, the behaviour of the jet becomes independent of $Re^\star$, and is as seen in Fig. 70.

Jet discharged into a confined, narrow channel

Figures 71 and 72 show the case in which a jet is discharged upwards from a nozzle (bottom) into a confined narrow channel. The surrounding fluid is entrained into the jet, as a part of the energy of the jet is imparted to it.

Figure 71 shows that, once the jet is deflected to the left by a disturbance (the right control nozzle is open and the left one closed here), a flow which makes up for the entrainment of fluid into the jet occurs through a narrow passage, so the pressure of the left-hand-side of the jet becomes low enough to make the jet attach on the left hand wall. Figure 72 shows the reverse case.

This is another example of the *Coanda effect* which causes a wall jet to remain attached to the wall. Using this effect, *fluidics* can switch the main jet by means of a small control jet, giving the effect of a switching element with amplification.

Figs. 66 and 67. Velocity distribution of a free jet.

Water, outlet velocity 4.3 cm/s, outlet width 11 mm, $Re = 520$, hydrogen bubble method.

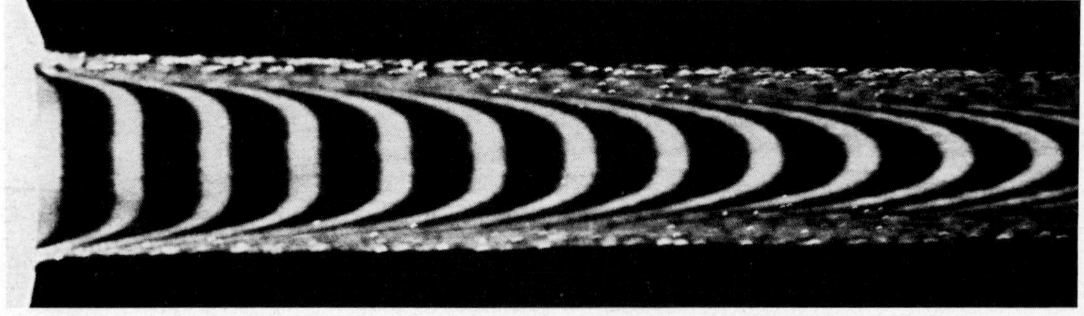

 Water, outlet velocity 9.2 cm/s, nozzle diameter 10 mm, $Re = 280$, hydrogen bubble method.

Fig. 68. Collision of two opposed jets (water, outlet velocity 1.3 cm/s, nozzle diameter 12 mm, $Re = 150$, hydrogen-bubble method).

Fig. 69. Axisymmetric underexpanded jet (air, Mach number 1.0, nozzle diameter 10 mm, $Re = 1.03 \times 10^6$, colour-schlieren method).

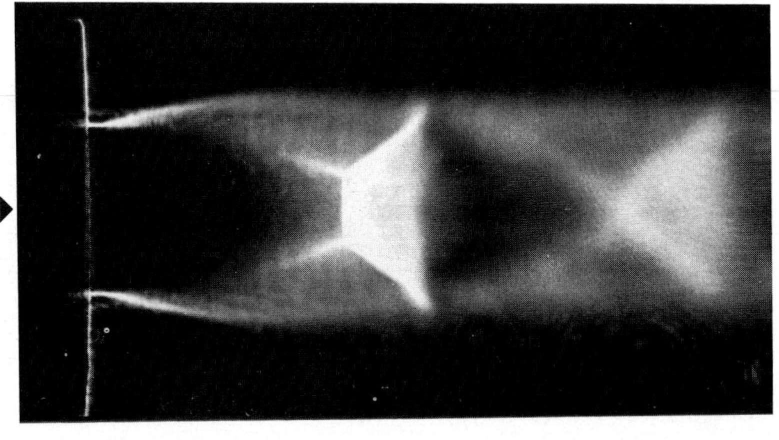

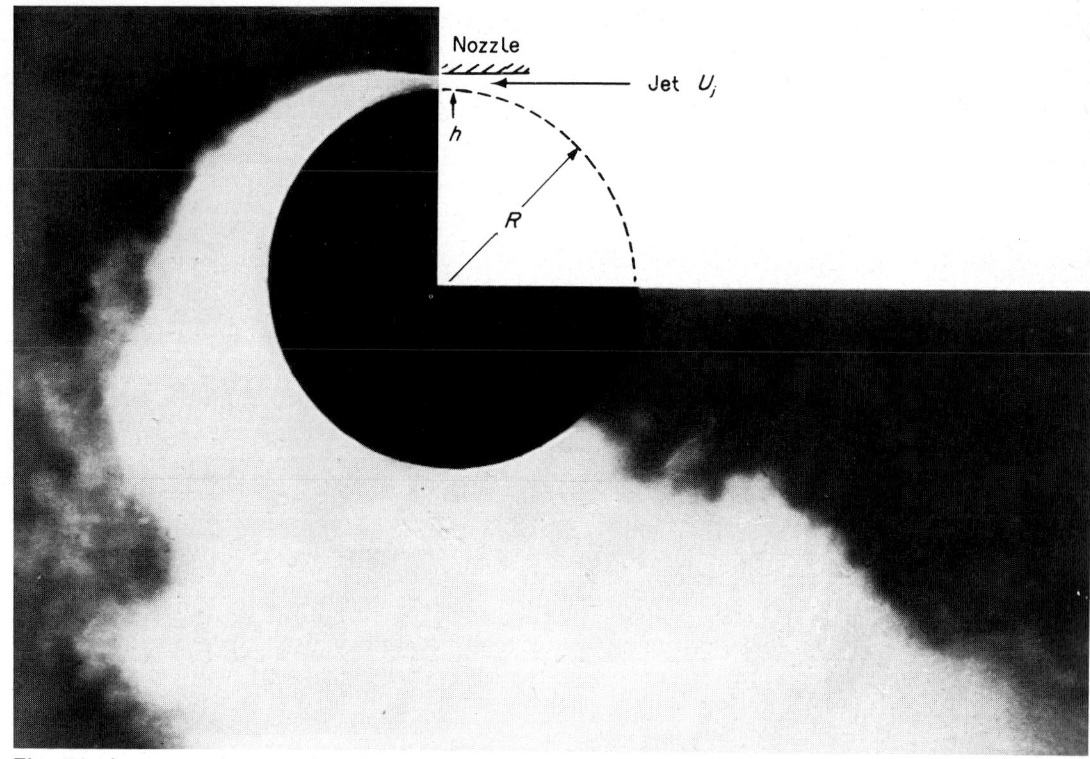

Fig. 70. Jet curved around a cylinder surface (dry ice fog and air, $U_j = 20$ m/s, $h = 3.175$ mm, $R = 50.8$ mm, $Re^* = 1.4 \times 10^4$).

Figs. 71 and 72. Jet discharged into a confined narrow channel.

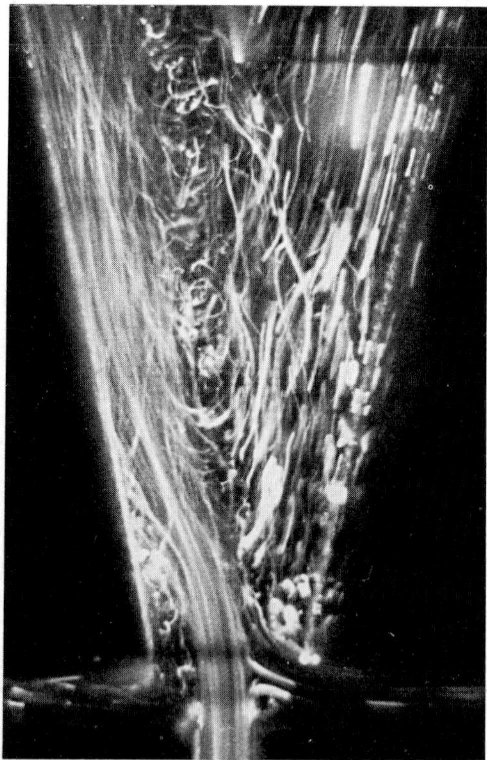

Fig. 71. Water, jet velocity 5.5 m/s, main nozzle width and depth 1 mm × 5 mm, *polystyrene-tracer method.*

Fig. 72. Water, jet velocity 0.9 m/s, main width and depth 5 mm × 5 mm, *polystyrene-tracer method.*

Miscellaneous Features of the Flow Past Solid Bodies: Wakes

Karman vortex streets behind circular cylinders

Karman vortices are generated behind a single circular cylinder immersed in the flow in a certain Reynolds number range, as described in the explanation of Figs. 3 and 4. The Karman vortices alternately departing from a single circular cylinder are shown in Fig. 73. Rhodamine B dye is injected into the flow from holes located at points ± 70° from the front stagnation point.

The interference between vortices behind two circular cylinders in tandem varies according to the ratio S/d and the Reynolds number, where S and d indicate the distance between the centres and the diameter of the cylinders, respectively. The Strouhal number $St = \lambda d/U$ of the rearward cylinder increases as S/d increases and asymptotically approaches the value for the single circular cylinder. Here, λ, d and U indicate the vortex discharge frequency, cylinder diameter and incident flow speed, respectively. The synchronized state of vortex generation behind the two cylinders in tandem is shown in Fig. 74.

Karman vortices behind circular cylinders side-by-side also interact with each other and make various flow patterns according to the value of S/d.

The case where the generation of Karman vortices has opposite phase between neighbouring cylinders is shown in Fig. 75. This is a typical flow pattern when the value of S/d is relatively high.

Wake behind single circular cylinder

A fairly long exposure time (1 sec.) shows the wake of a single circular cylinder in Fig. 76. This is the time-averaged wake envelope, though the actual wake is turbulent. If a shorter exposure time is used, the turbulence of the wake can be seen.

The white lines showing the envelope are visualized by fluoresceine dye injected from holes located at points ± 30° from the front stagnation point. The flow in the wake is visualized by Rhodamine B dye injected from 5 holes located at points ± 120 – 180°.

Smoke is used to show the flow around a circular cylinder on a flat plate in Fig. 77. Besides the general shape of the wake, a horseshoe or 'collar' vortex can be seen around the cylinder in the boundary layer on the plate. A *collar vortex* is usually generated when a protuberance is placed in a boundary layer.

Wakes of two circular cylinders

The wakes of two circular cylinders side-by-side interact with each other and make various flow patterns according to the value of the S/d ratio, where S and d are the distance between the cylinders and the diameter, respectively.

Fig. 73. Karman vortex street behind single circular cylinder.

Fig. 74. Karman vortex streets behind circular cylinders in tandem.

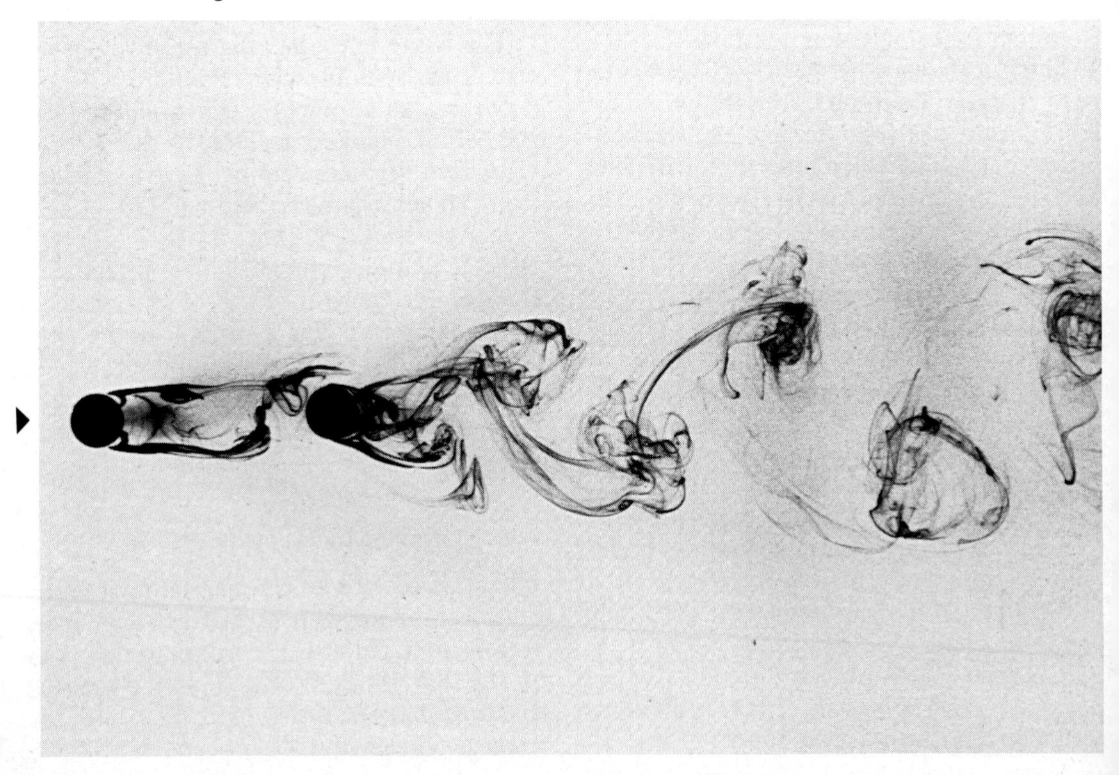

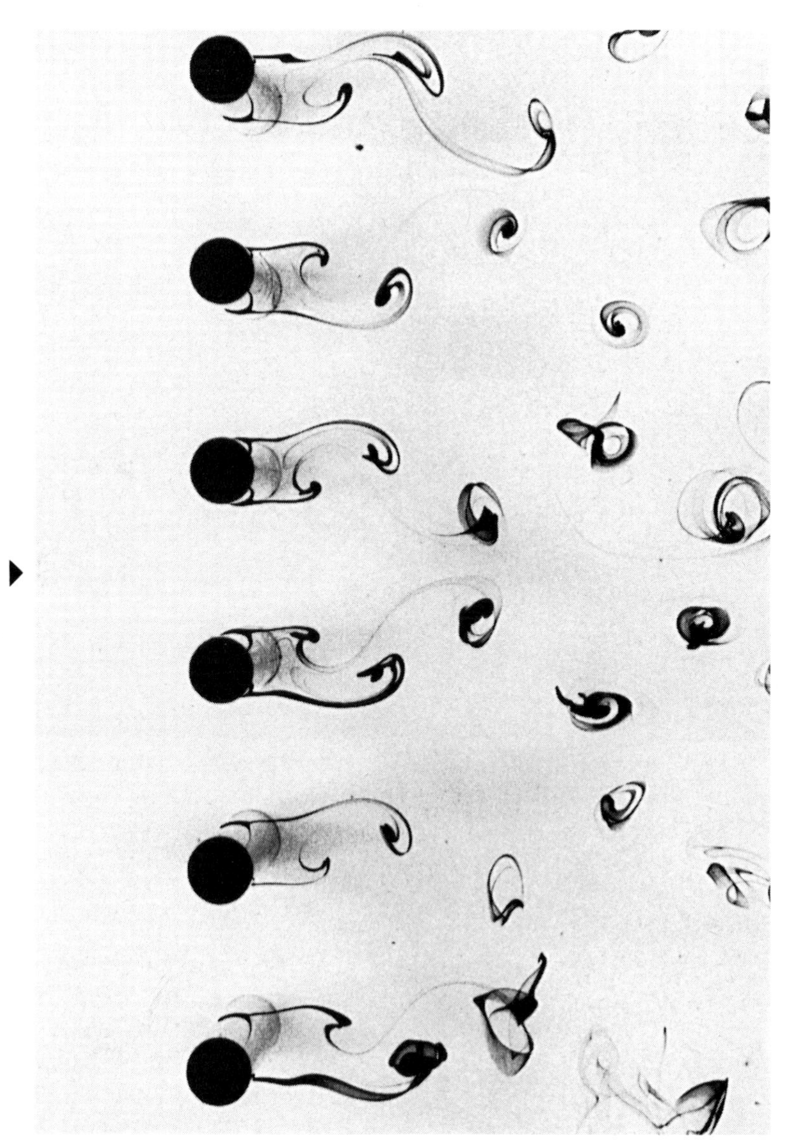

Fig. 75. Karman vortex streets behind circular cylinders side-by-side [water, flow speed 1.4 cm/s, cylinder diameter 15 mm, centre distance 75 mm (74), 45 mm (75), injected streak line method].

The case of the relatively high value of $S/d = 3.0$ is shown in Fig. 78. Here the wakes of the cylinder scarcely interact with each other and each is virtually the same as that of a single cylinder as shown in Fig. 76.

On the other hand, as the distance between the cylinders decreases, an asymmetric flow pattern develops because of increasing interference between the wakes, as shown in Fig. 79 ($S/d = 1.5$).

Three-dimensional structure of the wake behind a circular cylinder of finite length

The flow field around a cylindrical body of finite length, placed transverse to the flow direction, is markedly dependent on the length of the cylinder when the cross-section is blunt and there is a separated flow region. The drag, the fluctuation in lift, and the

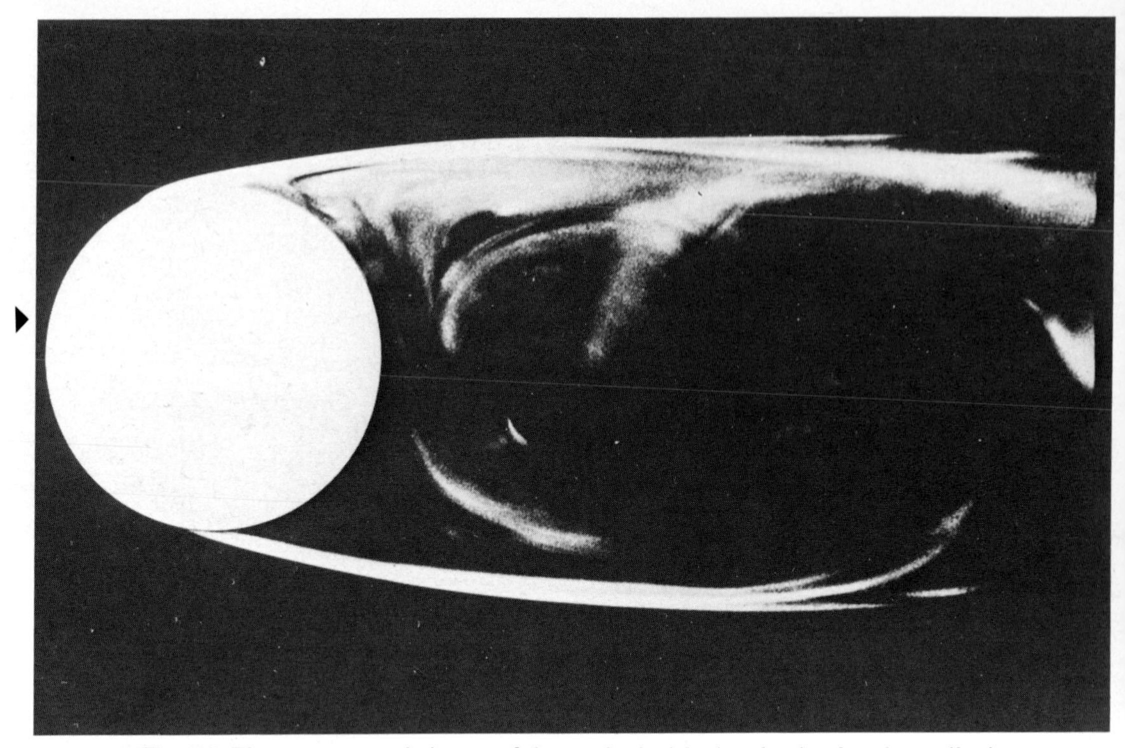

Fig. 76. Time-averaged shape of the wake behind a single circular cylinder (water, flow speed 8.3 cm/s, cylinder diameter 15 mm, $Re = 1.24 \times 10^3$, injected streak line method).

Fig. 77. Horseshoe or 'collar' vortex around a circular cylinder
(air, flow speed 50 cm/s, cylinder diameter 88 mm, $Re = 2.9 \times 10^3$, injected streak line method, liquid paraffin mist method).

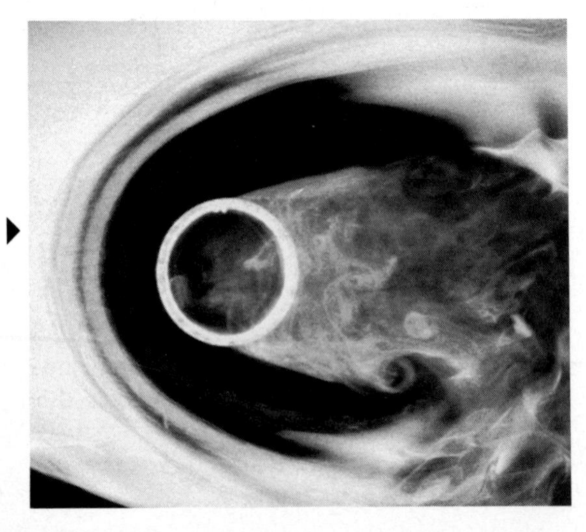

discharge frequency of vortices, all vary greatly from the corresponding two-dimensional values.

Part of the incident flow enters the separation region over the end of the body, as shown in Fig. 83(b). This causes a rise in the pressure of the separation region compared with that of the two-dimensional case.

The wake of a circular cylinder with a finite length is shown in Figs. 80 and 81. It is visualized by a dye injected in the flow from holes on the surface of the cylinder at intervals of 2 mm.

Cross-sectional views from a downstream view point are shown in Fig. 82. It is seen that the vortex released from the cylinder is deformed as it moves away downwind. The parameter X/d is the distance downstream from the cylinder normalized by the cylinder diameter.

Side views of a single filament of dye flowing around the end of a cylinder are shown in Fig. 83. The dye is injected at the point $Z/d = 0$ in the upper view and $Z/d = 0.5$ in the lower view, respectively. Here, Z

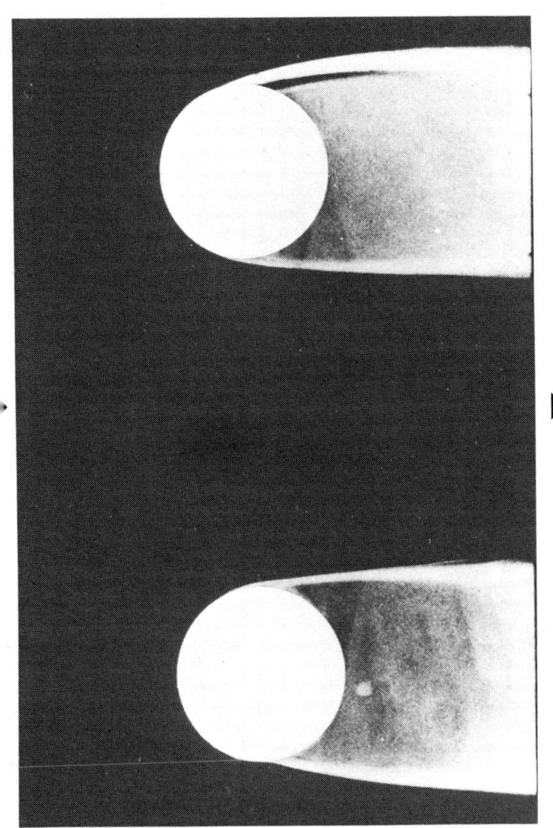

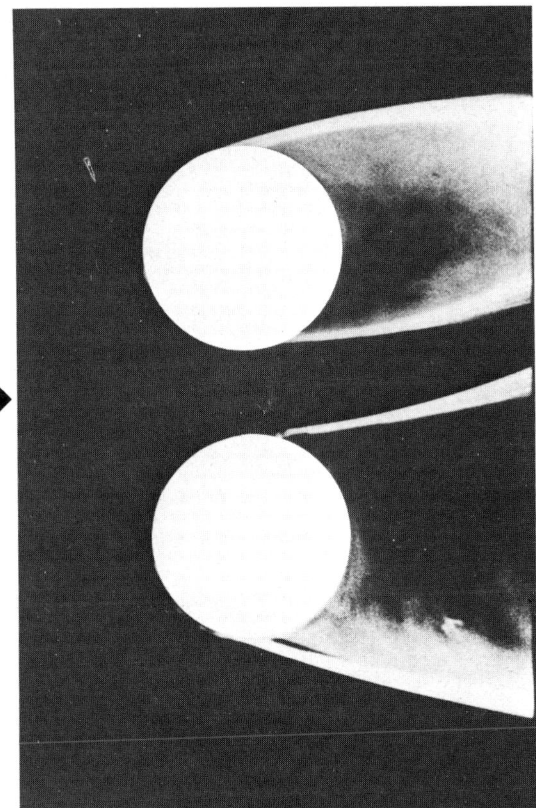

Figs. 78 and 79. Interference of wakes behind two circular cylinders side-by-side (water, flow speed 6.3 cm/s, cylinder diameter 15 mm, $Re = 900$, injected streak line method).

indicates the distance from the end of the cylinder.

In Fig. 83(a) the dye does not enter the wake. On the other hand, Fig. 83(b) indicates that part of the incident flow going over the end of the cylinder enters deep into the wake, reaches the rear stagnation region of the cylinder, and then flows downstream with the discharged vortices.

Cross sectional views of pairs of eddies in the neighbourhood of the end of the cylinder are shown in Fig. 84. The pair of eddies is stable and stationary when $Z/d < 1.5$. As the value of Z/d increases, a region is reached where the eddies are unstable and extend and shrink by turns as shown in Fig. 84(b). At still larger values of Z/d, the eddies are periodically discharged downstream.

It is concluded that the wake of a circular cylinder with finite length is different from a two-dimensional one, and because part of the incident flow enters the wake around the end of the cylinder, has the following characteristics:

(1) There is a pair of eddies in the vicinity of the end of the cylinder, and these are almost stationary.

(2) Further from the end of the cylinder, vortices are shed which have a strong three-dimensional structure.

Flow around the surface of a circular cylinder (limiting stream lines)

An oil film surface flow pattern, produced by two-dimensional flow over the surface of a circular cylinder, is shown in Fig. 85. The oil flow clearly shows the laminar separation, reattachment, transition to turbulence and turbulent separation. A marked cross-flow develops as the laminar separation is approached because of the gravity.

However, the flow pattern changes as the Reynolds number changes. When the Reynolds number is much lower for example, laminar separation occurs but re-attachment and turbulent separation do not.

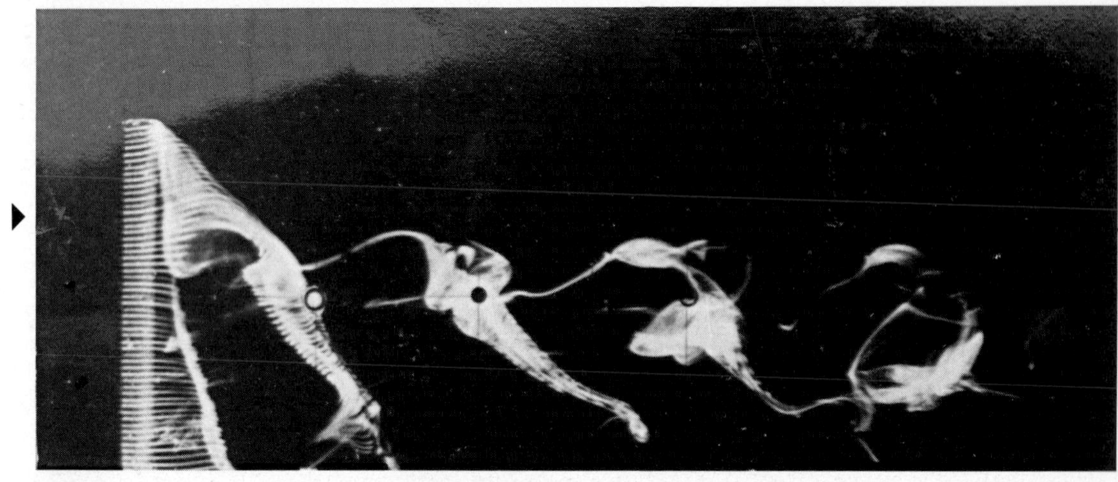

Fig. 80. Side view of a circular cylinder of finite length.

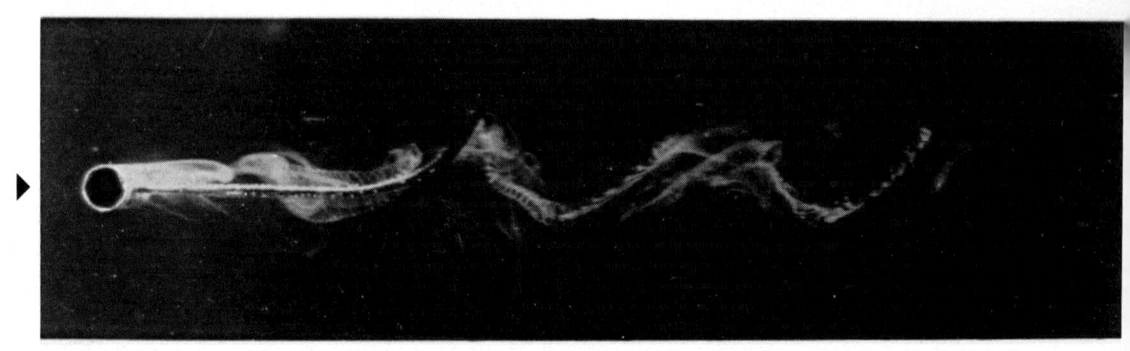

Fig. 81. Top view of a circular cylinder of finite length.

Fig. 82. Rear view of a circular cylinder of finite length.

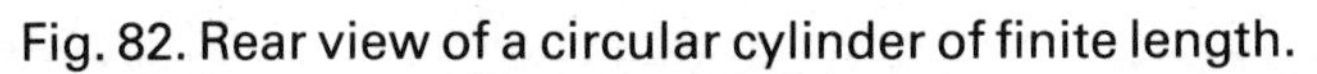

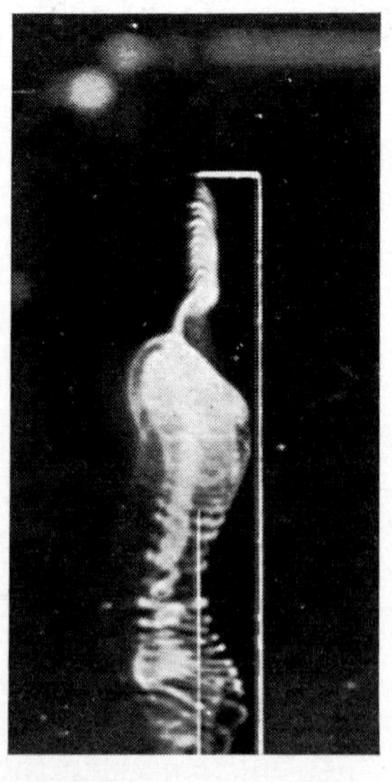

(a) $X/d = 5.3$

(b) $X/d = 10$

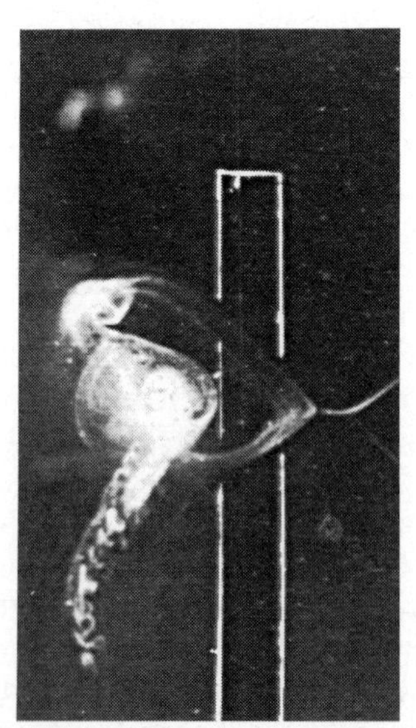

(c) $X/d = 16$

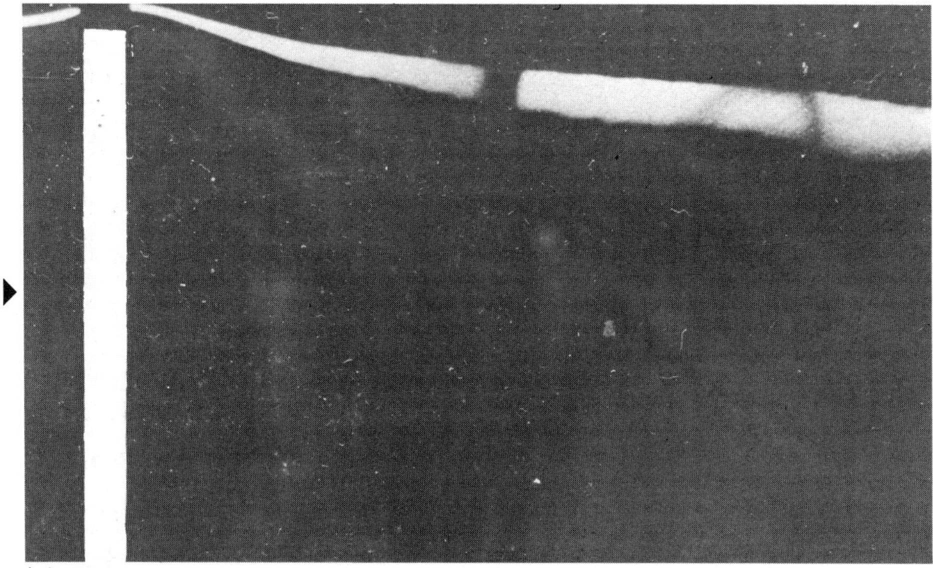

(a) $Z/d = 0$

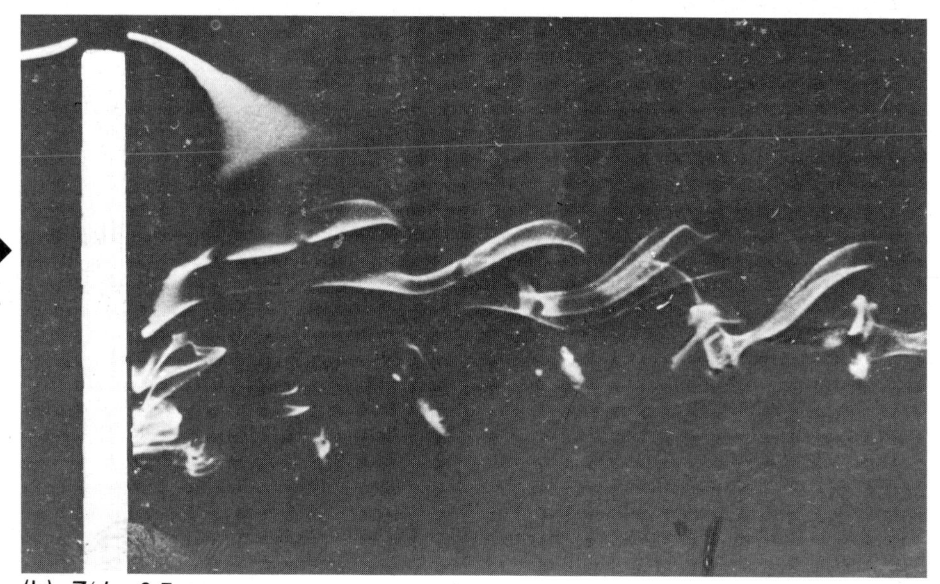

(b) $Z/d = 0.5$

Fig. 83. Side view of the wake near the end of a vertical circular cylinder of finite length (water, flow speed 1.26 m/s, cylinder diameter 11 mm, $Re = 120$, injected streak line method).

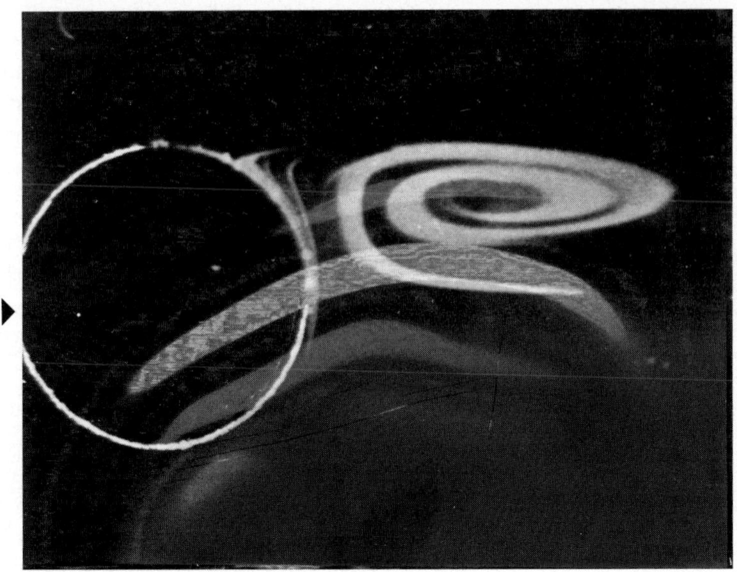

(a) $Z/d = 1.4$
(b) $Z/d = 2.2$

Fig. 84. Cross-sectional view of the wake near the end of a circular cylinder of finite length (water, flow speed 1.26 m/s, cylinder diameter 11 mm, Re = 120, injected streak line method).

Wake of a falling sphere

The wakes of spheres falling through fluid are shown in Figs. 86 and 87. The sphere accelerates immediately after it is released, but eventually falls at a constant speed when gravity minus buoyancy balances the drag force. The drag coefficient is a function of the Reynolds number. A case in the range of Reynolds number where Stokes formular $C_D = 24/Re$ holds, and only a very steady, diffuse and completely laminar wake exists is shown in Fig. 86. On the other hand, the wake of the sphere is intense and highly turbulent at higher Reynolds number as shown in Fig. 87.

Wake of a flat plate

When a flat plate is placed parallel to the incident flow, the boundary layer does not separate until it reaches the trailing edge. There is no periodic discharge of vortices at the trailing edge, but the wake becomes unstable as it flows downstream, and a vortex

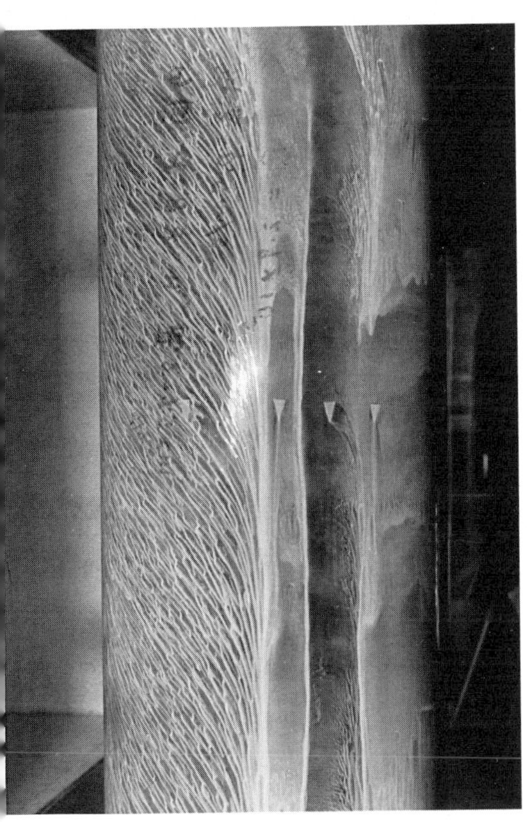

Fig. 85. Flow along surface of a circular cylinder (air, flow speed 10 m/s, $Re = 4 \times 10^5$, oil flow method).

Fig. 86. Glycerine, flow speed 0.61 m/s, sphere diameter 25.4 mm, $Re = 2.5 \times 10^{-1}$, dye painting method.

Fig. 87. Water, flow speed 4.7 m/s, sphere diameter 25.4 mm, $Re = 1.2 \times 10^4$, dye painting method.

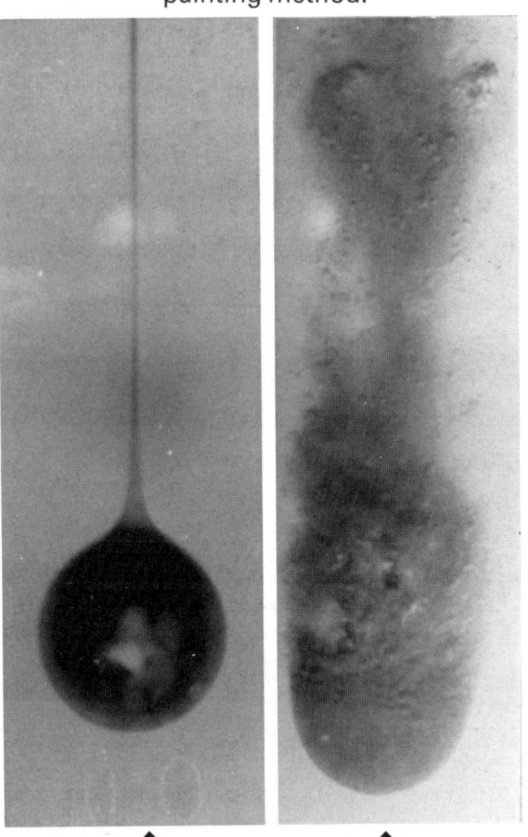

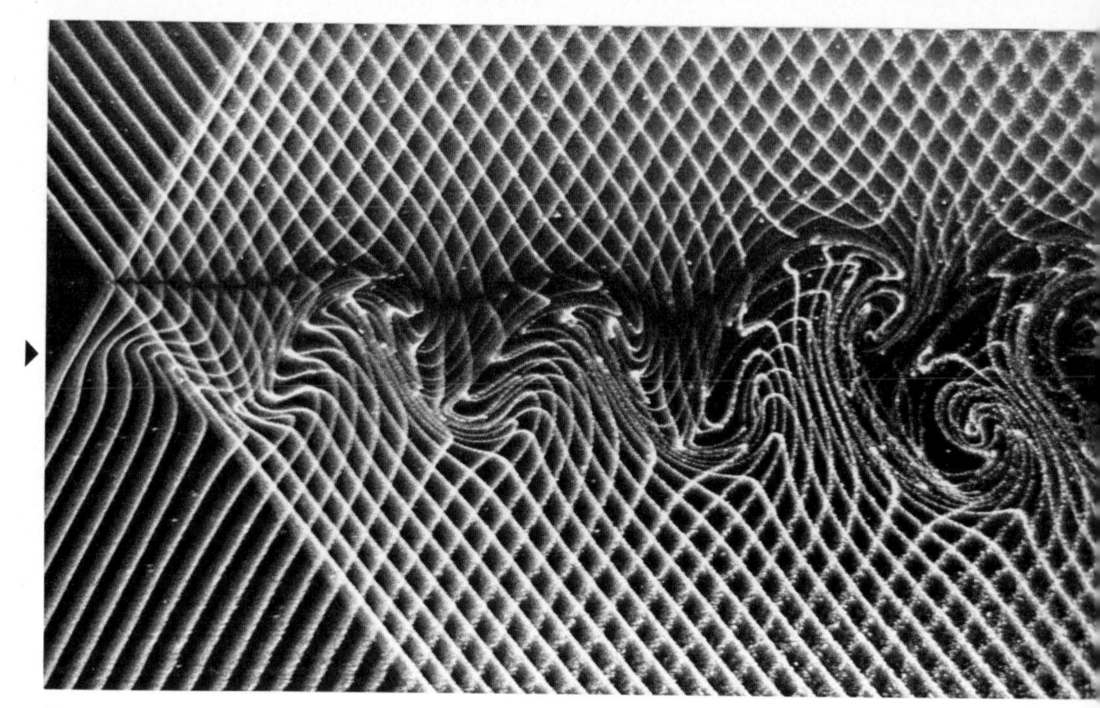

Fig. 88. Water, flow speed 2.2 cm/s, plate length 50 mm, Re = 900, hydrogen bubble method.

Fig. 89. Water, flow speed 3.4 cm/s, wing thickness 8 mm, Re = 280, hydrogen bubble method.

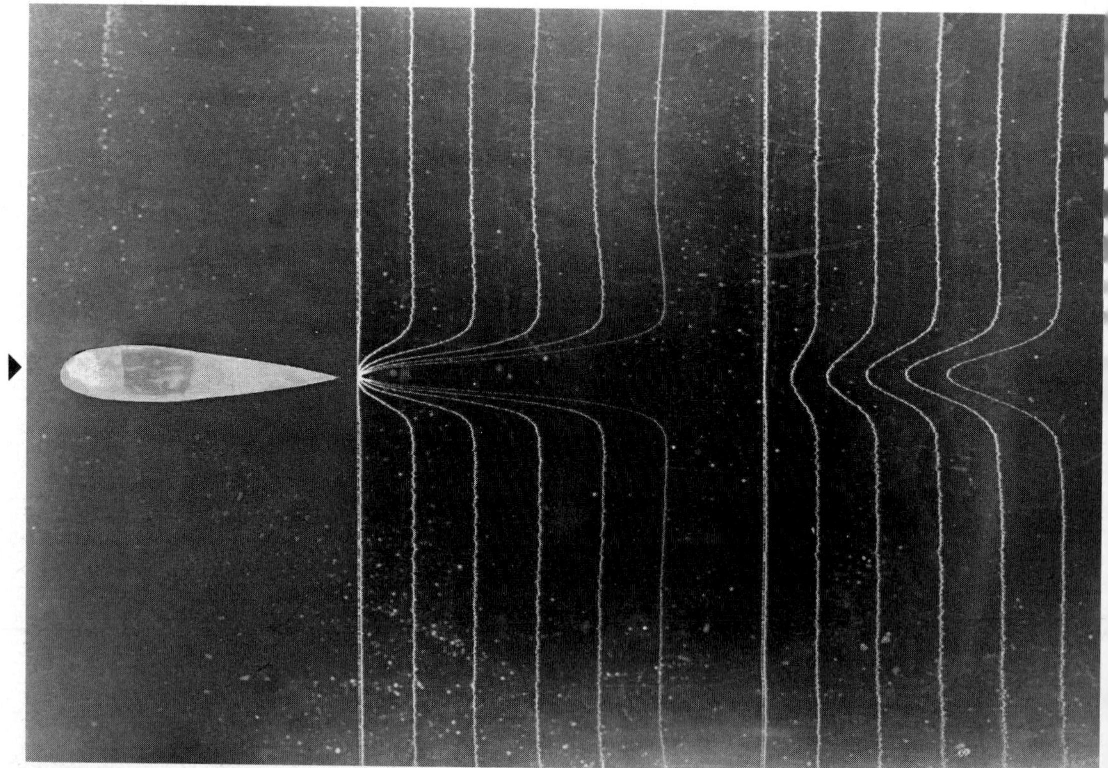

street is gradually formed. The vortex street is similar to the Karman vortex street and has a similar periodic arrangement. Such a delayed vortex street formation is also observed in the wake of a circular cylinder with an extremely low Reynolds number.

Time lines of small hydrogen bubbles are shown in Fig. 88. They are generated by pulses from X-shaped filaments placed downstream of the plate. The deformation and drift of diamond-shaped fluid elements can be clearly seen in the time line meshes.

Wake of an aerofoil

In the case of a body with a sharp trailing edge such as a wing, the boundary layers formed along the upper and lower surfaces unite to form a wake. As the wake passes downstream, the minimum air speed in it increases, and its width increases as it spreads and diffuses. This is because the fluid particles in the wake take on momentum from the main flow.

Time lines delineated by small hydrogen bubbles generated from two filaments placed near the trailing edge and perpendicular to the incident flow are shown in Figure 89. Wake diffusion and reduction in intensity can be clearly seen in this figure.

Wake of a delta wing

The flow separates and vortices are generated at the leading edge of a flat delta wing. These vortices are called leading *edge separation vortices.* They become a pair of *longitudinal vortices* as they move downstream.

Leading edge separation vortices of a flat delta wing are shown in Fig. 90. There are a pair of longitudinal vortices and a *down wash* in the centre. The length and interval of the tufts are 50 mm and 10 mm, respectively.

Longitudinal vortices at the stern of a ship

Flow separates from a three-dimensional body to form a complicated shape often rolling up into *longitudinal vortices.*

Longitudinal vortices in the wake of a full model ship of ordinary shape are shown in Fig. 91. The water line is approximately located at the top of the figure, and the bottom at the third grid step. Also white lines showing cross sections of the ship are drawn at intervals of 5% of the overall length.

A large pair of longitudinal vortices is observed which is symmetrical to the centre plane of the hull. These vortices increase drag. They also gather up the wake of the hull boundary layer and this greatly affects the propulsive performance of propellers at the stern.

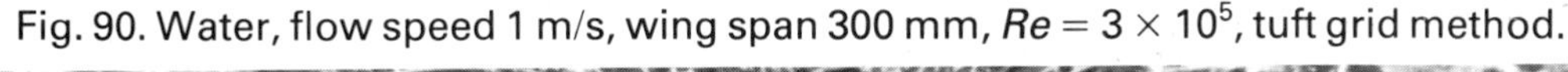

Fig. 90. Water, flow speed 1 m/s, wing span 300 mm, $Re = 3 \times 10^5$, tuft grid method.

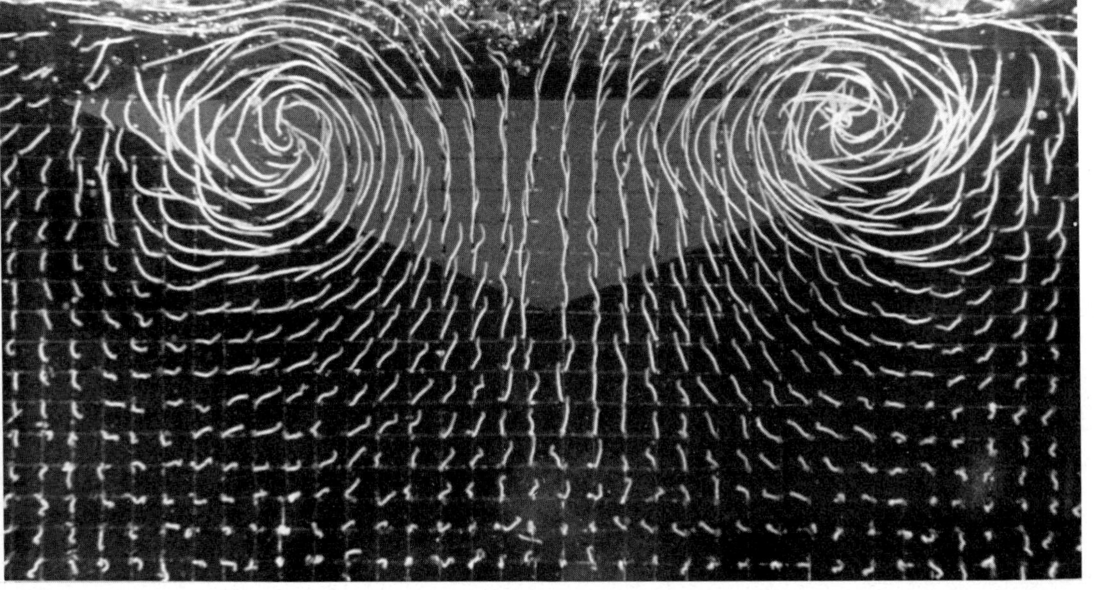

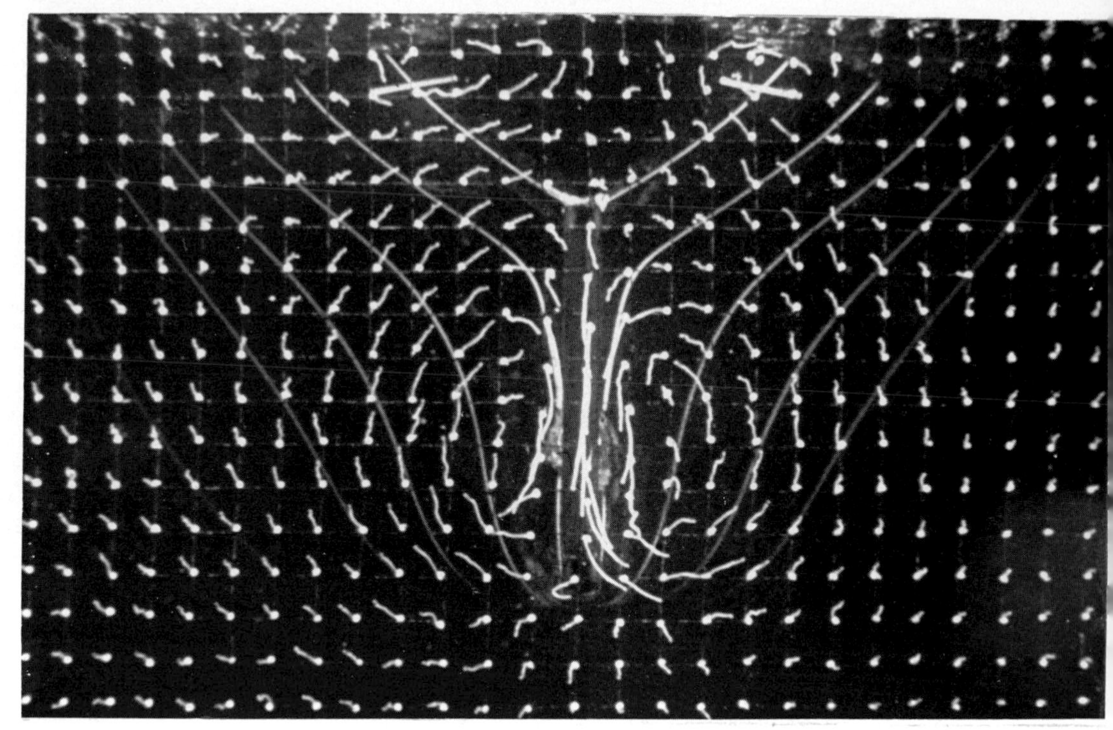

Fig. 91. Water, flow speed 0.8 m/s, ship length 2.0 m, $Re = 1.6 \times 10^6$, tuft grid method.

In addition, longitudinal vortices are generated from the bow, the bottom of the bow, and the stern near the water line. However, they are not clear in Fig. 91.

The length and separation of the tufts are 50 and 10 mm, respectively.

Limiting streamlines on an axisymmetric ellipsoid

This oil flow pattern approximately shows the *limiting streamlines* on an axisymmetric ellipsoid. They are known to vary according to the ratio of the length to the width and the angle of attack of the ellipsoid.

The case of an angle of attack of 15° is shown in Fig. 92. Studs to generate turbulence are attached on the upstream end of the model.

The region on the rear half of the model where the oil does not flow is a bubble-type separation. The bunching of oil like a belt immediately upstream of the bubble, shows a three-dimensional separation which becomes a pair of *longitudinal vortices* similar to those shown in Figs. 90 and 91.

Limiting streamlines on three-dimensional bodies

The boundary layer on a three-dimensional and non-axisymmetric body has a distorted velocity distribution and its mode of separation is different from the two-dimensional case.

The *limiting streamlines* around the bottom near the conventional bow of a full ship model are shown in Fig. 93. The experiment was conducted in a recirculating water channel and observed from below. The lines on the hull surface are drawn at intervals of 5% of the overall length. The flow along the side of the ship goes toward the bottom but never crosses a certain line. This line forms an envelope around the oil flow lines and is a three-dimensional separation line.

The limiting streamlines in the stern region of a conventional ship model are shown in Fig. 94. The flow goes from the bottom to the side as it flows rearwards, but it never flows upward of a certain line. Again this line is a three-dimensional separation line. The separated flow forms a pair of *longitudinal vortices* as shown in Fig. 91.

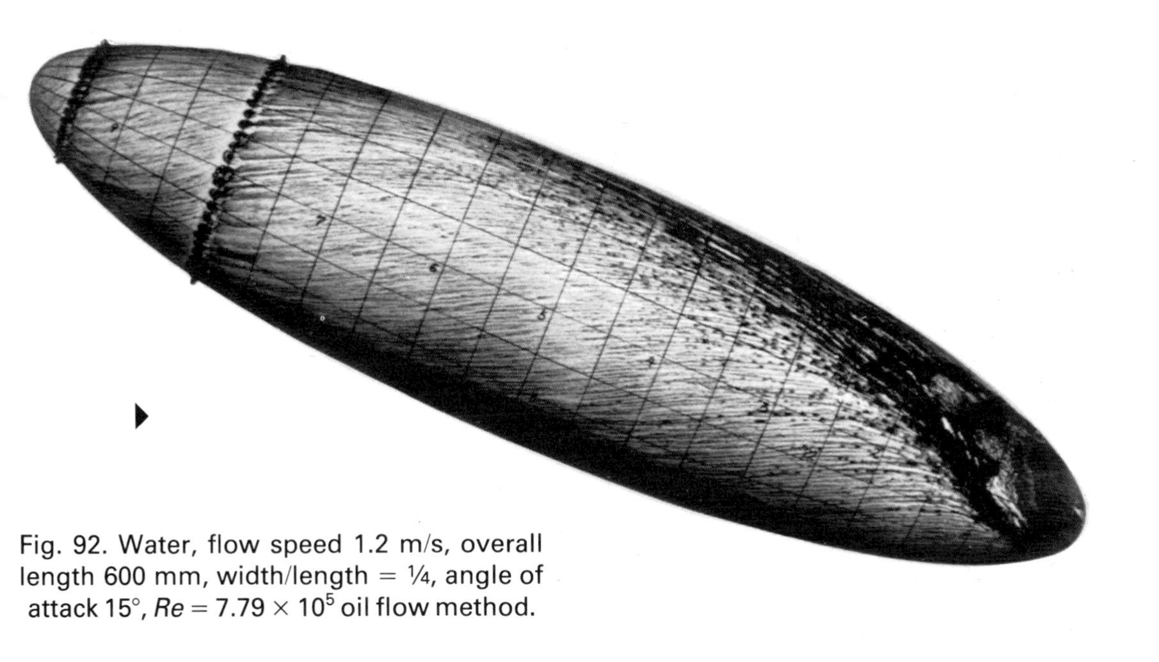

Fig. 92. Water, flow speed 1.2 m/s, overall length 600 mm, width/length = ¼, angle of attack 15°, $Re = 7.79 \times 10^5$ oil flow method.

Fig. 93. Flow near the bow of a model ship.

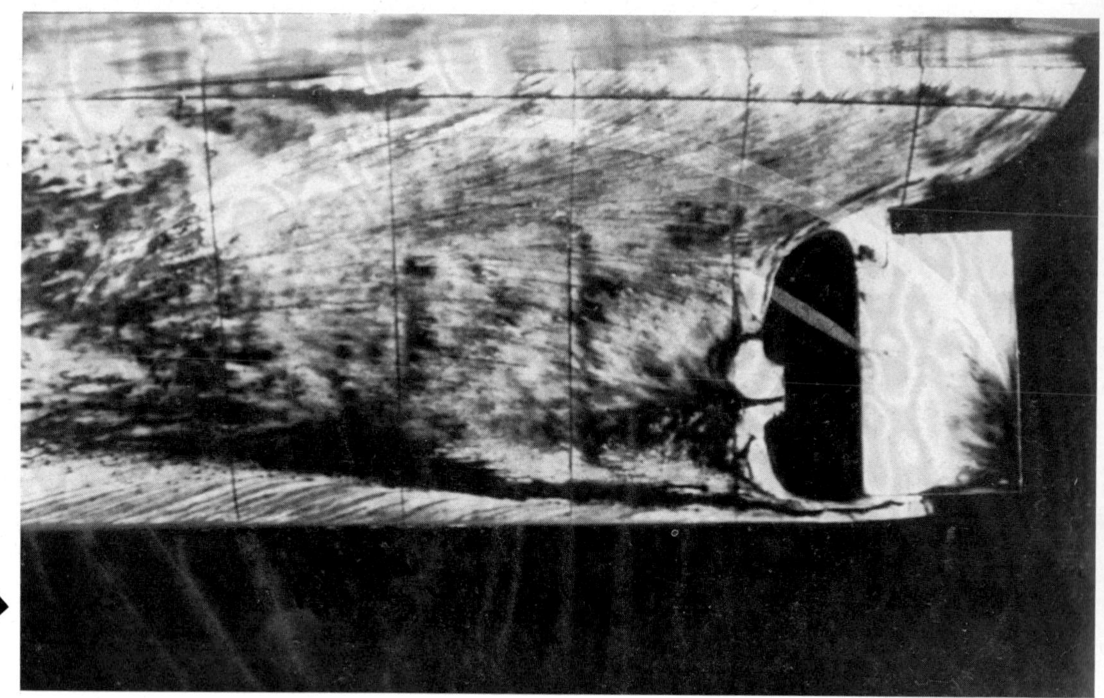

Fig. 94. Flow near the stern of a model ship
(water, flow speed 0.8 m/s, overall length 2000 mm, $Re = 1.6 \times 10^6$, oil flow method).

More than one three-dimensional separation line and corresponding pairs of longitudinal vortices may be generated when the shape of the stern is complicated.

Flow around a car model

Generally speaking, because a car body is not very aerodynamic, many separated regions of various sizes are generated at the front of the hood and roof, the rear half of the body, etc.

The flow around car models is visualized in a smoke tunnel in Figs. 95 and 96. A bluff model is shown in Fig. 95. Here it is apparent that the flow separated from the front of the hood collides with the wind shield. Such separated flow increases aerodynamic drag. The *drag coefficient* C_D of this model is 0.85.

Fig. 95. Flow around a bluff car model.

Fig. 96. Flow around a streamlined car model
(air, flow speed 4 m/s, wheel base 500 mm, $Re = 1.3 \times 10^5$, three-dimensional smoke tunnel).

The flow field around a streamlined sports car model is shown in Fig. 96. Here, the flow field is more refined than that of the former case, because no separation occurs at the front of the hood and roof, and the wake region is narrow. The value of C_D for this particular model is 0.32, much lower than the average value for passenger cars (0.4).

Wake of a car body

The wake of a car body varies according to the shape of its rear half. With a fast back car as in the upper left diagram in Fig. 97, the flows separating from the rear pillars form a pair of *longitudinal vortices* and induce the flow from the roof to remain attached over the rear window. On the other hand, the upper right illustration is a typical example of a sedan with bubble-type separation generated on the trunk.

The flow field of a fastback is also shown in Fig. 97. The pair of the longitudinal vortices in the wake and the resulting induced *downwash* are apparent. It is well known that lift and drag increase considerably when this type of the flow occurs.

The length and separation of the tufts are 50 and 10 mm, respectively.

Flow around skyscrapers

Figure 98 shows the flow around a skyscraper, made visible by the use of smoke from titanium tetrachloride painted on the windward walls and the roofs of the model building. The model is of a skyscraper 250 m in height in the sub heart of Tokyo, and its scale is 1/750.

The wind deflected by the windward wall blows over the building or around the side, and separation occurs from the sharp edges. Separated flow occurring on the side of a building is the main cause of wind damage. Separations on the top and sides are apparent in Fig. 98.

Flow around buildings arranged in an H shape

Flow directions near the ground around three buildings which are arranged in an H shape are shown in Fig. 99. Small flags made of foaming polystyrol are used at intervals of 25 mm. The flags are designed to rotate smoothly on beads threaded on brass pipes.

The flow field is very complicated when a few buildings are nearby. Macroscopically, the wind takes a path round the buildings, as if they were as a single object. On a smaller scale, the pressure distributions made by the buildings interact with each other and govern the flow field among the buildings.

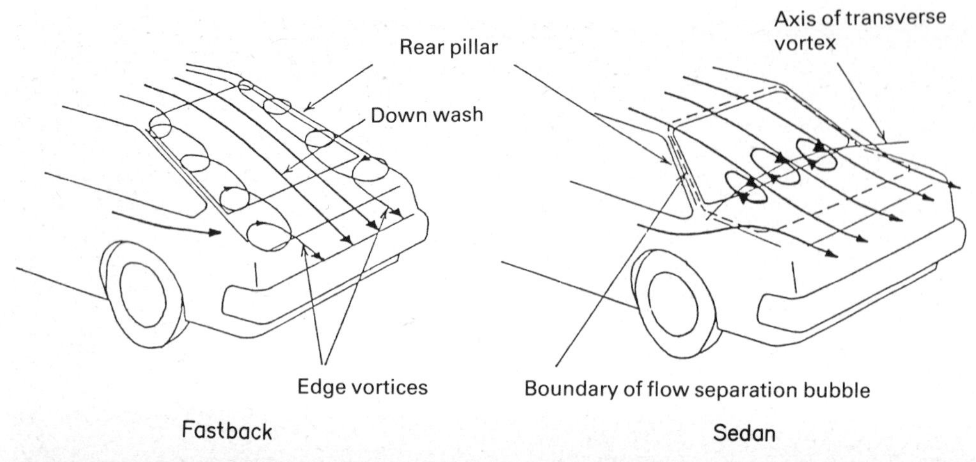

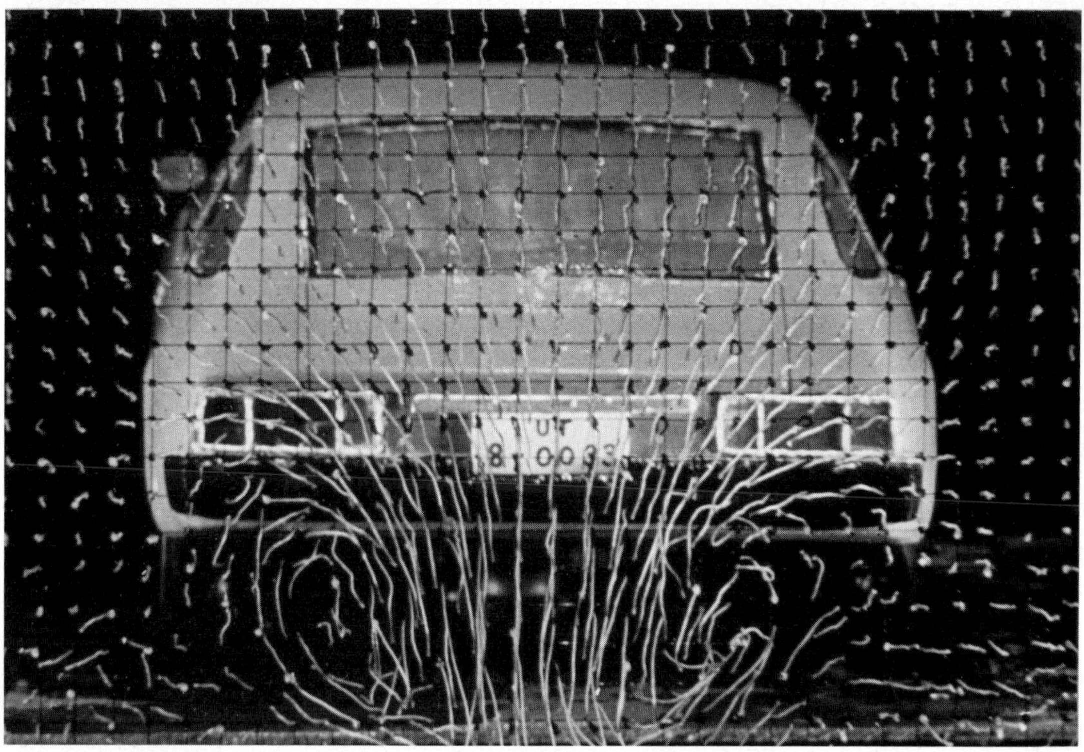

Fig. 97. Flow field of a fastback car (water, flow speed 1 m/s, overall length 530 mm (1/8-scale), $Re = 5 \times 10^5$, tuft grid method).

Fig. 98. Air, flow speed 0.5 m/s, width of skyscraper 100 mm, $Re = 3.4 \times 10^3$, titanium tetrachloride method.

Fig. 99. Air, flow speed 6 m/s, width of building 100 mm, $Re = 4.2 \times 10^4$, depth-tuft, method.

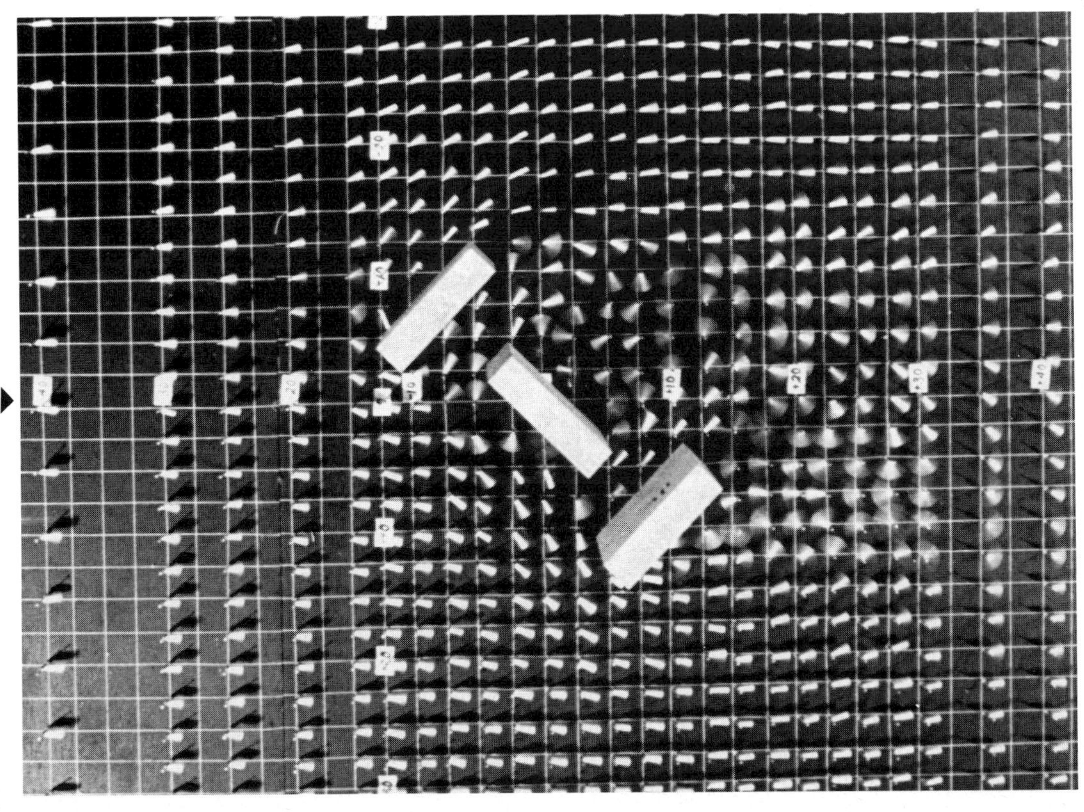

Pipe and Channel Flows

Flow at a pipe inlet

When fluid flows from a large vessel into a pipeline, streamlines entering the pipe form a *contraction* at the inlet. If the inlet is sharp-edged, the flow cannot follow the sudden change of direction of the surface and consequently the flow separates (Fig. 100). Some distance downstream, the flow reattaches to the wall and fills the pipe.

Energy loss in a flow element is generated mostly by velocity deceleration, being proportional to the square of velocity difference between its upstream and downstream sections. For the element shown in Fig. 100 the deceleration occurs just after *separation* at the inlet, and brings about large energy loss. However, when the pipe inlet is rounded as in Fig. 101, the energy loss is reduced notably, as there is no sudden change of flow direction, and accordingly no separation and no deceleration at the pipe inlet. When the pipe is located obliquely to the inflow, the separation is found only at one side (Fig. 102).

Inflow to a pipe inlet projecting into a vessel

When fluid flows from a large vessel with the pipe inlet projecting into it, *separation* occurs just after the inlet edge as in the case of Fig. 100, if the edge is left as cut (Fig. 103). With increasing flow *contraction* at the pipe inlet, the separation as well as the subsequent deceleration of flow are intensified and consequently the *energy* loss is increased. The *energy loss* in the case of Fig. 103 is larger than in the case of Fig. 100.

To prevent separation at a pipe inlet projecting into a large vessel, the inlet edge should be rounded with a larger radius than in Fig. 101. A similar radius to that in Fig. 101 is shown in Fig. 104.

Flow in a divergent pipe and near a pipe exit

In a *diffuser* (a duct with divergent walls) the pressure increases in the direction of flow. The fluid flowing near the wall loses energy due to wall friction, and accordingly, the velocity near the wall decreases with increasing distance from the entrance as shown in Fig. 105. At some point downstream, a *separation point* is reached, here the velocity gradient at the wall becomes zero. At the sections downstream of this point the velocity gradient at the wall is negative and reversed flow adjacent to the walls is observed.

When the separation is formed the effective section of flow passage is reduced, the pressure recovery which would otherwise be realised due to deceleration, is not achieved, and a large pressure loss is incurred. The maximum divergence angle with no separation is five to six degrees.

At a pipe exit as shown in Fig. 106, the flow separates at the edge of the exit, and decelerates gradually to come to a standstill. The kinetic energy at the exit does not diffuse to increase the pressure, but is dissipated to thermal internal energy. When the exit is rounded as in Fig. 107, the separation is graded down and the loss can be reduced slightly. To decrease the energy loss it is preferable to provide the exit with a diffuser with an angle of six to eight degrees.

Flow through a turning passage

Velocity in a turning passage varies from a streamline to streamline. It is inversely

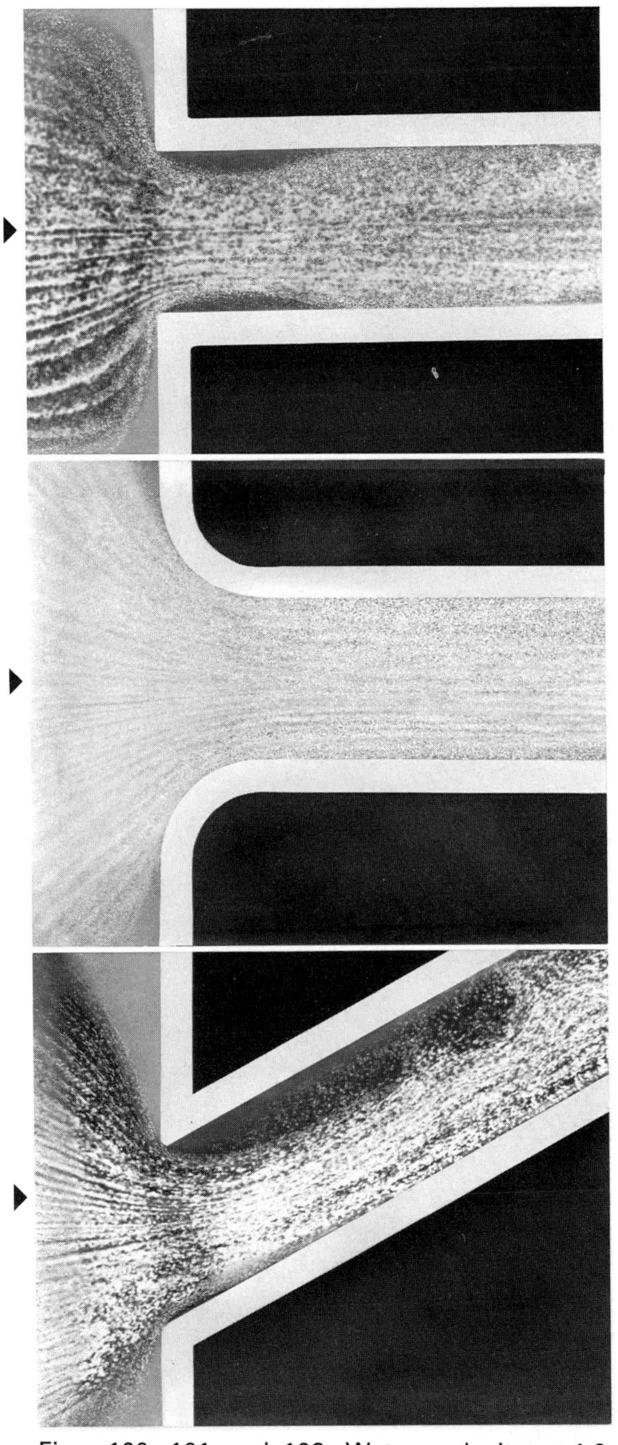

Figs. 100, 101 and 102. Water, velocity = 1.2 cm/s, passage width = 20 mm, *Re* 240, hydrogen bubble method.

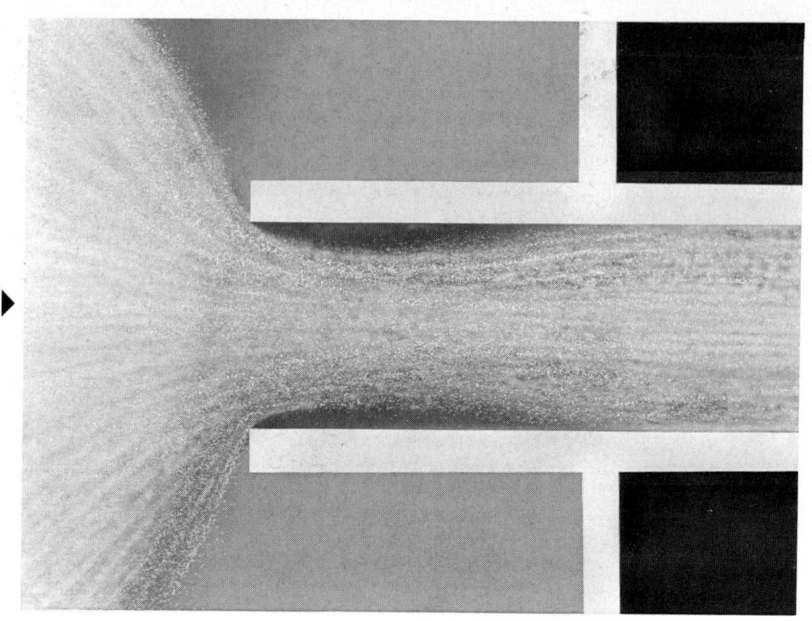

Fig. 103. Water, velocity = 3.4 cm/s, passage width = 20 mm, $Re = 500$, hydrogen bubble method.

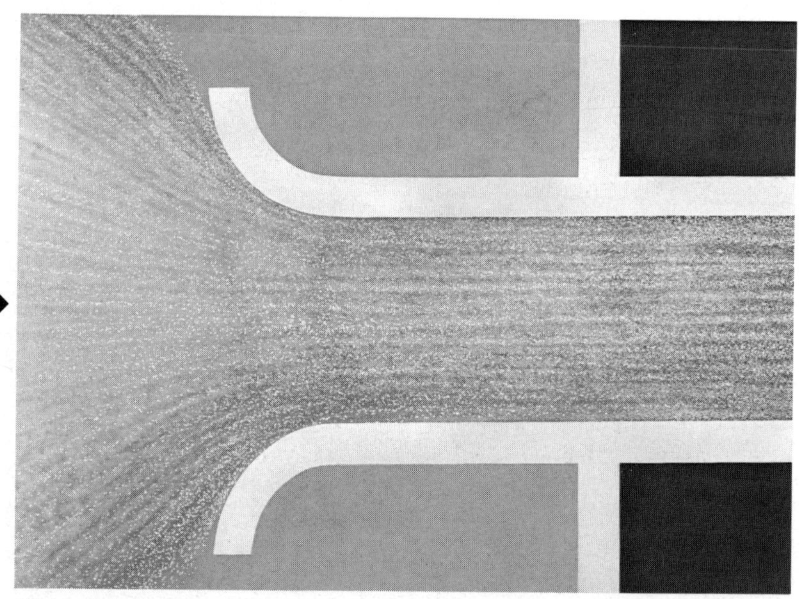

Fig. 104. Water, velocity = 3.4 cm/s, passage width 20 mm, $Re = 600$, hydrogen bubble method.

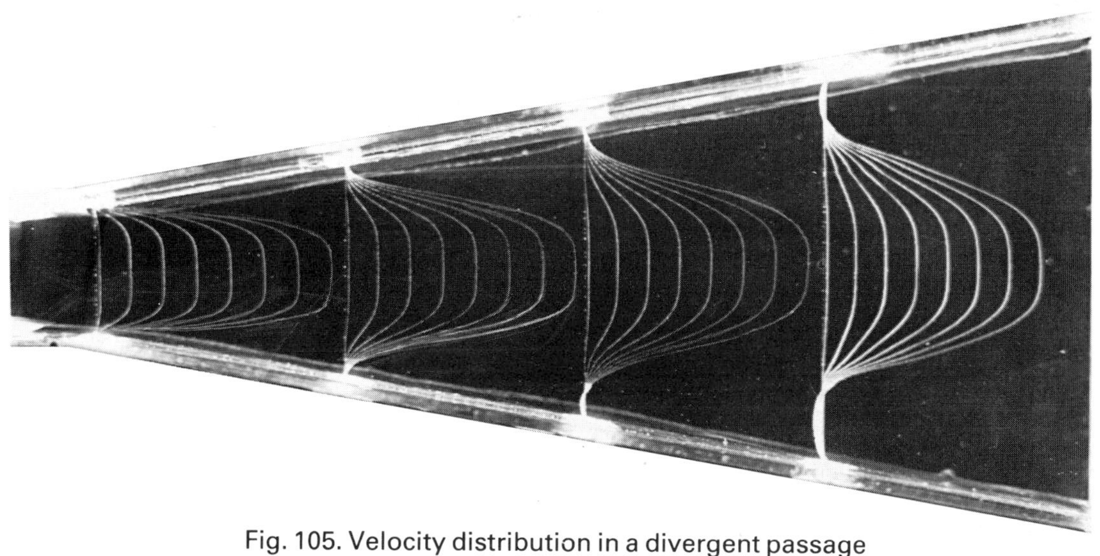

Fig. 105. Velocity distribution in a divergent passage (water, velocity = 3.0 cm/s, inlet width = 30 mm, divergence angle = 20°, Re = 900, hydrogen bubble method).

proportional to the radius of curvature, and shows higher values with decreasing radius. Accordingly, the pressure decreases with decreasing radius. The streamlines, however, run concentrically, as the pressure gradient – gradient inward – is balanced with the centrifugal force – directed outward.

After the turning passage the flow enters the straight again with a uniform section. Here the velocity should become uniformly distributed. Here, the inner streamlines have to be decelerated and separation from the wall occurs as shown in Fig. 108 under the same condition as mentioned for Fig. 105. After a certain distance the separated region disappears and flow runs full in the passage, where the velocity shows sudden decrease and the energy loss is generated. With decreasing channel width to radius ratio, the separated region, and the corresponding energy loss become smaller. But it should be noted that the energy loss due to friction increases with decreasing width to radius ratio.

When *guide vanes* as shown in Fig. 109 are provided in the turning portion, the separated region and thus the loss due to the turning decrease markedly, as the width to radius ratio for individual passages becomes small.

Figs. 106 and 107. Flow at exit from passages (water, velocity = 7.6 cm/s, passages width = 20 mm, Re 1.5×10^3, hydrogen bubble method).

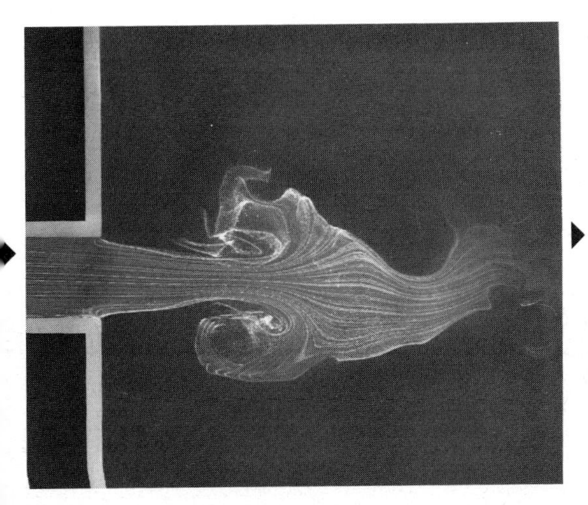

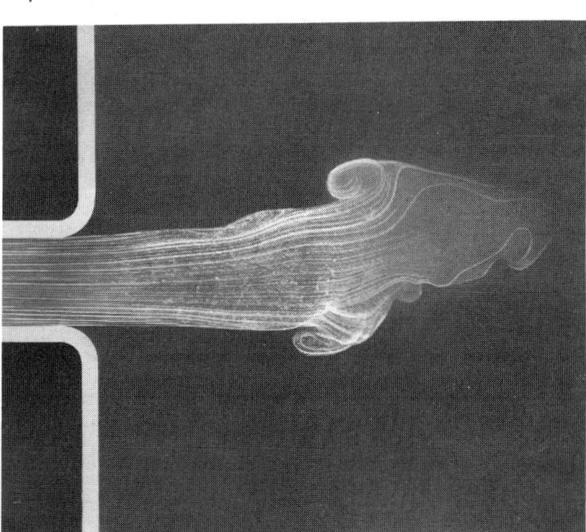

Flow through a bend and elbow

A piping element with a small radius of curvature compared with its width is called an elbow, whereas that with a large radius is called bend. In the case of Fig. 110 two passages are connected directly as a mitred joint. The velocity of each streamline is proportional to its curvature radius. Therefore, the velocity at the inner corner is theoretically infinite. In reality, however, the flow separates there. On the other hand, at the outer corner a region with higher pressure is generated owing to the flow running against the corner, thus forming a stagnation region.

With an elbow the velocity at the inner corner, the separation and *consequently* the energy loss are much larger than with a bend. The loss can be reduced markedly by providing *guide vanes* at the corner region. This is shown in Fig. 111.

Flow in sudden contraction and enlargement sections

With the flow in a *sudden contraction* (Fig. 112), a stagnation region is observed at the corner just ahead of the contraction, as the pressure on the wall increases against the corner. In the passage after the contraction the flow is similar to that at the pipe inlet, and the flow contraction and the separation are generated. Then, along with disappearance of the separation the flow runs full in the downstream pipe. The energy loss is caused mainly by velocity decrease due to the disappearance of the separation after the contraction.

The flow in a *sudden enlargement* (Fig. 113) produces a *separation* at the inlet corner of the downstream pipe, but soon runs full in the downstream pipe. This sudden decrease of velocity brings about a large *energy loss*.

Flow in a diffuser with a large divergence

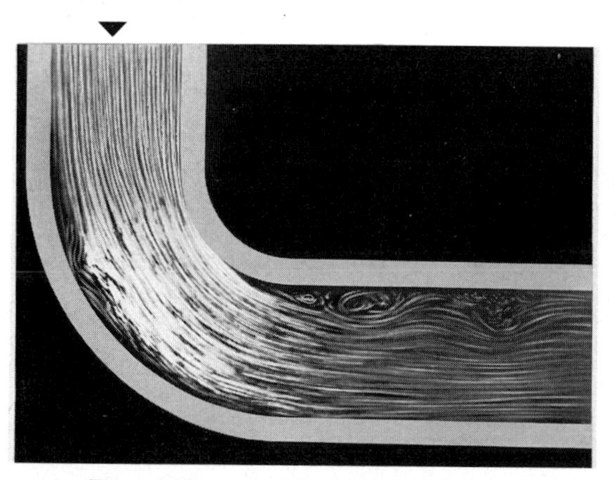

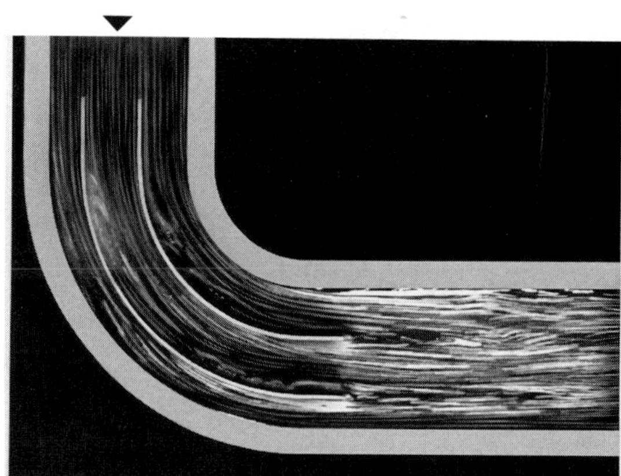

Figs. 108 and 109. Flow through a turning passage with and without guide vanes (water, velocity = 10 cm/s, passage width = 20 mm, $Re = 2 \times 10^3$, surface tracer method).

Figs. 110 and 111. Passages connected in right angle with and without guide vanes (water, velocity = 10 cm/s, passage width = 20 mm, $Re = 2 \times 10^3$, surface tracer method).

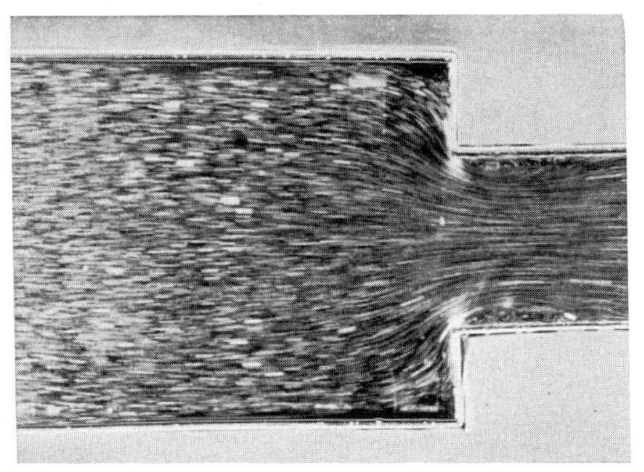

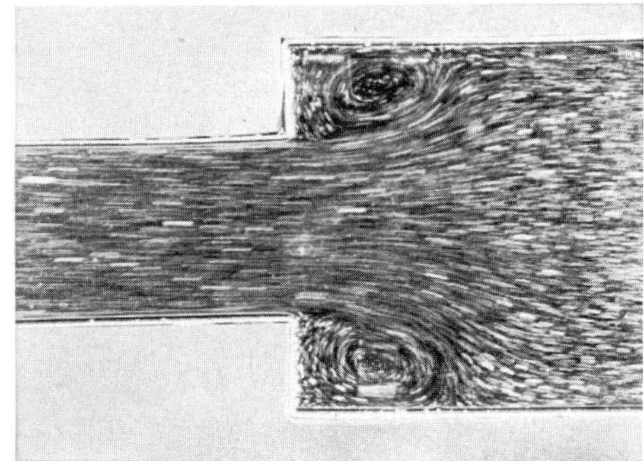

Figs. 112 and 113. Flow through sudden contraction and enlargement (water, velocity = 15 cm/s (112), 20 cm/s (113), width of narrower passage = 400 mm, tracer method).

Figs. 114 and 115. Slow and fast diverging flow (water, velocity = 3 ~ 4 cm/s (114), 15 ~ 20 cm/s (115), width of narrower passage = 600 mm, diverging angle = 20°, tracer method).

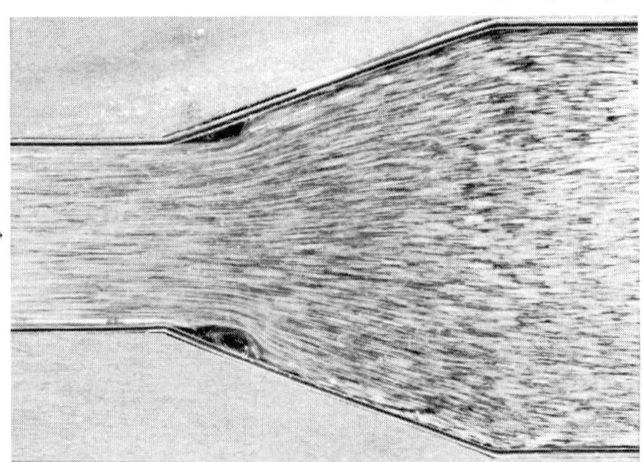

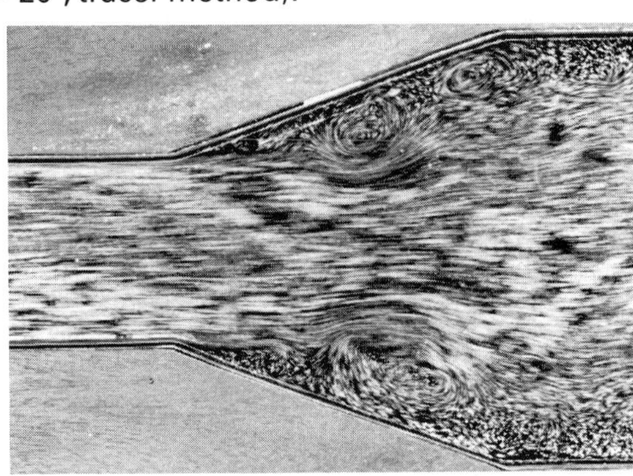

angle is accompanied by separation on both sides of the passage. When the velocity , and hence the Reynolds number, is very low (Fig. 114), the separation region is small in scale, and the flow reattaches rapidly and runs full in the passage, as the viscous forces on the fluid predominate over the inertia forces.

On the contrary, when the velocity, and hence also the Reynolds number, is high (this is the usual case), the separation is formed on both sides, and the energy loss caused is almost as great as in the case of a sudden enlargement of section.

Flow through an orifice and a nozzle

In Fig. 116 the nozzle has a curved upstream surface converging smoothly up to the nozzle opening, thus contracting the flow to a diameter equal to the nozzle diameter.

As shown in Fig. 117, the flow through a sharp-edged orifice continues to contract after it emerges and the diameter of the contracted flow from the orifice is smaller than that from a nozzle of the same size.

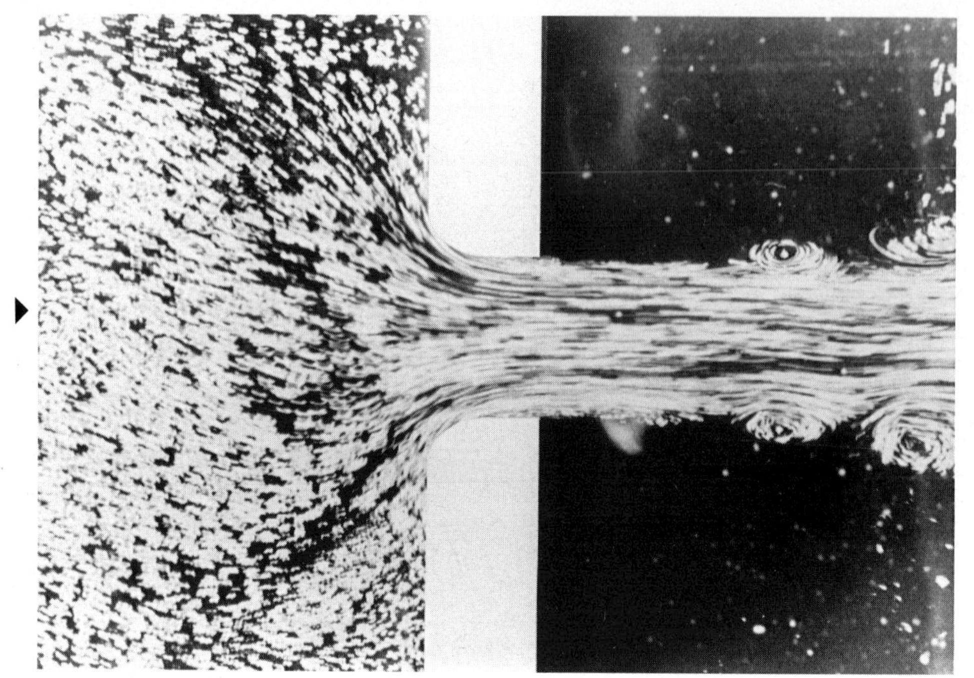

Fig. 116. Flow through a nozzle.

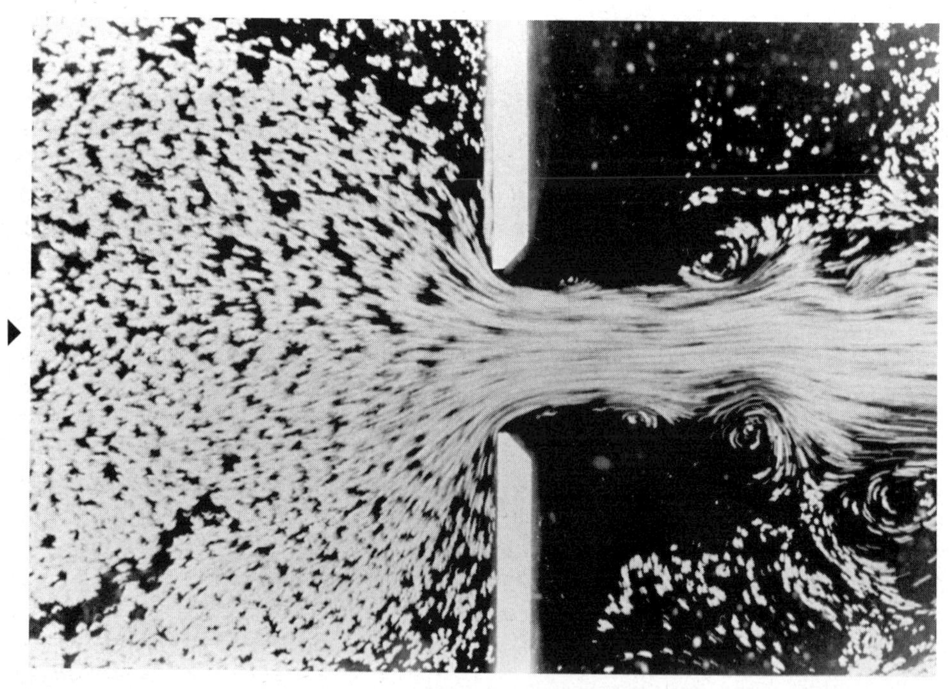

Fig. 117. Flow through an orifice
(water, velocity = 14 cm/s, width of opening = 30 mm, $Re = 4.3 \times 10^3$, hydrogen bubble method).

Circulation

Circulation theory of lift

The amount of lift force acting on an aerofoil section placed in a uniform flow can be evaluated using circulation theory. This was formulated independently by Kutta in 1902 and Joukowski in 1906.
The *Kutta–Joukowski theorem* is expressed by

lift per unit span (L) = density of fluid (ρ)
+ velocity of uniform flow (U)
+ circulation (Γ)

The theorem can be illustrated by utilizing the flow around a rotating circular cylinder. Shown in Fig. 118 is the flow around a stationary cylinder with smoke streak lines arranged symmetrically with respect to a horizontal plane through the centre of the cylinder. The arrangement of smoke lines shows that temporal mean lift force is zero.

In the case of the cylinder rotating in a clockwise direction shown in Fig. 119, the smoke lines are closely spaced above the cylinder (corresponding to higher velocity, lower pressure) and widely spaced below it (corresponding to lower velocity, higher pressure). As a result, the cylinder experiences an upward lift force normal to the undisturbed flow.

The fact that a rotating object that is moving in a fluid develops a fluid-dynamic side force has been known since early days. It was named *Magnus effect* and is quoted frequently to explain the curved flight of a spinning ball.

Circulation around an aerofoil section

The existence of circulation around a rotating cylinder is easily understood because the fluid particles adjacent to the cylinder move with the surface under the effect of viscosity. But the existence of circulation around an aerofoil at constant incidence is not so easily recognized. Evidence of this circulation may be obtained by visualization of the motion of fluid around an aerofoil set in motion from rest. Figure 120 shows an aerofoil at rest and aluminium powder floating on the surface of water also at rest. Just after the aerofoil section has been set in motion there remains a *starting vortex* (with anticlockwise rotation) behind the aerofoil as shown in Fig. 121. On the other hand, Kelvin's theorem shows that the circulation along any closed curve drawn in the flow field of a frictionless fluid is constant with respect to time. As the circulation along any curve which contains the aerofoil section in Fig. 120 is zero, it must be so along the corresponding curve (elements of which consists of the same particles of fluid as in Fig. 120) in Fig. 121. Around the aerofoil in motion there must be a circulation of which the magnitude is the same as, and the direction of rotation is opposite to (clockwise), that of the starting vortex. In a steady flow the magnitude of the circulation around an aerofoil section is so determined that the fluid particles travelling along upper and lower surfaces leave the trailing edge with the same velocity, and also have the same pressure. This effect is associated with the viscosity of the fluid and is called the *Kutta condition*. In an ideal fluid, when a thin uncambered aerofoil of chord length c is placed at an angle of attack α in a uniform flow of velocity U, the magnitude of circulation required to satisfy the Kutta condition can be shown to be

$$\Gamma = \pi U c \sin\alpha \approx \pi U c \alpha \qquad (1)$$

The lift force per unit span is

$$L = \rho U \Gamma = \pi \rho U^2 c \alpha$$

and the lift coefficient is

$$C_L = \mathrm{L}/(\rho U^2 c/2) = 2\pi\alpha \qquad (2)$$

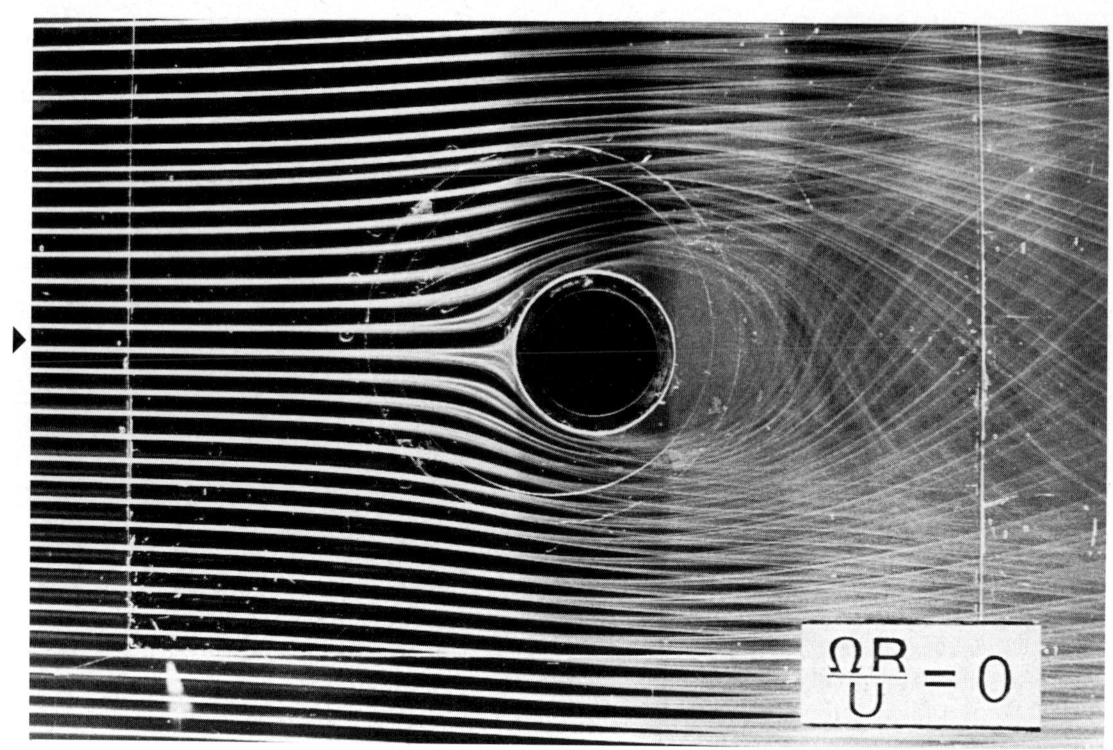

Fig. 118. Flow around a stationary circular cylinder
(air, $U = 4$ m/s, radius of cylinder $R = 72$ mm, $Re = 4 \times 10^4$, smoke streak line method).

Fig. 119. Flow around a rotating cylinder
(air, $U = 4$ m/s, $R = 72$ mm, $\Omega R/U = 4$, $Re = 4 \times 10^4$).

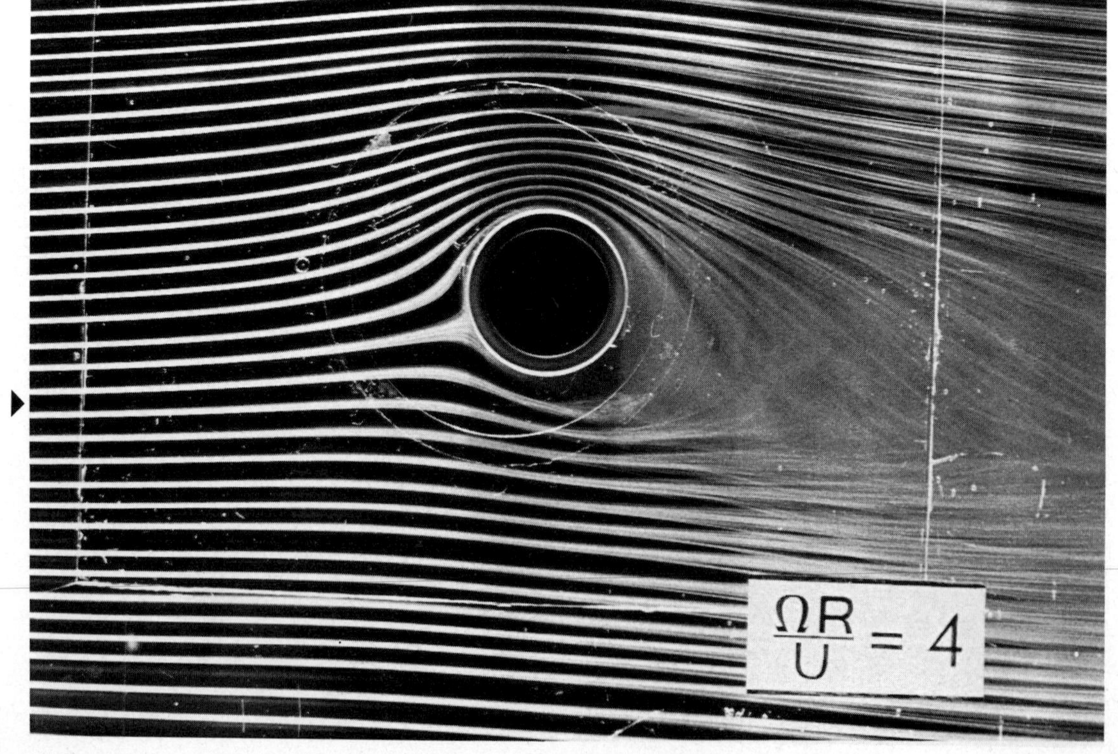

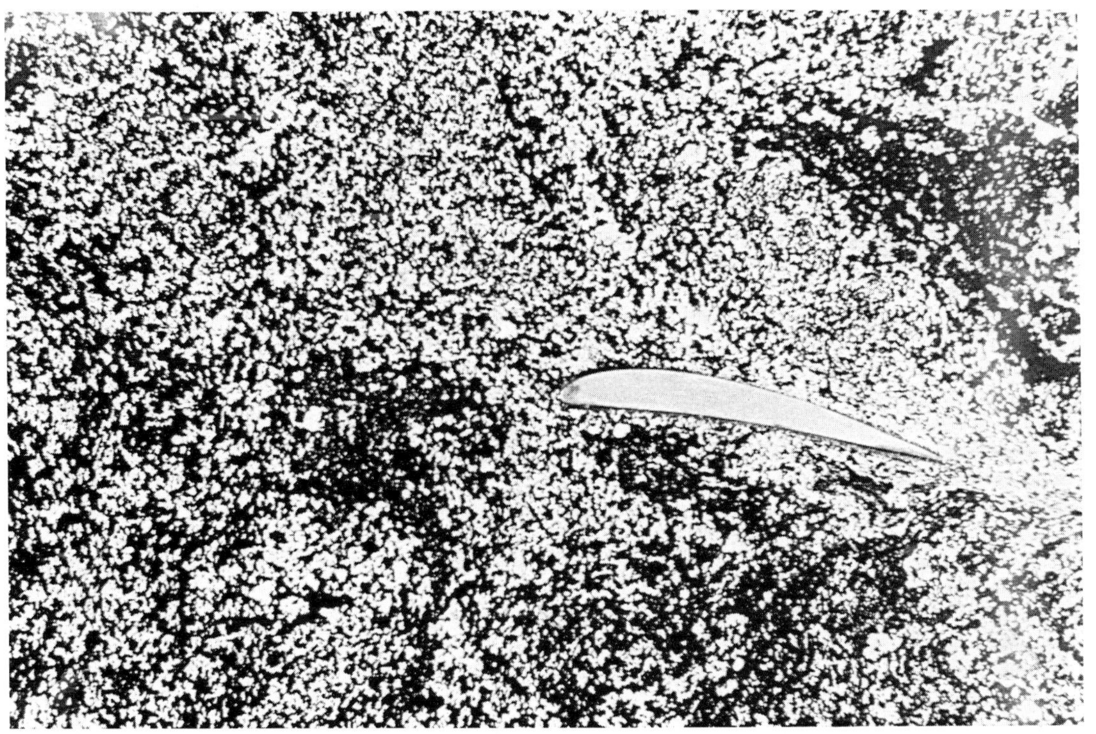

Fig. 120. Aerofoil section in rest
(water, $U = 0$ m/s, chord length $c = 180$ mm, floating tracer method).

Fig. 121. Starting vortex behind an aerofoil section
($U = 30$ cm/s, $c = 180$ mm, $Re = 5 \times 10^5$, floating tracer method).

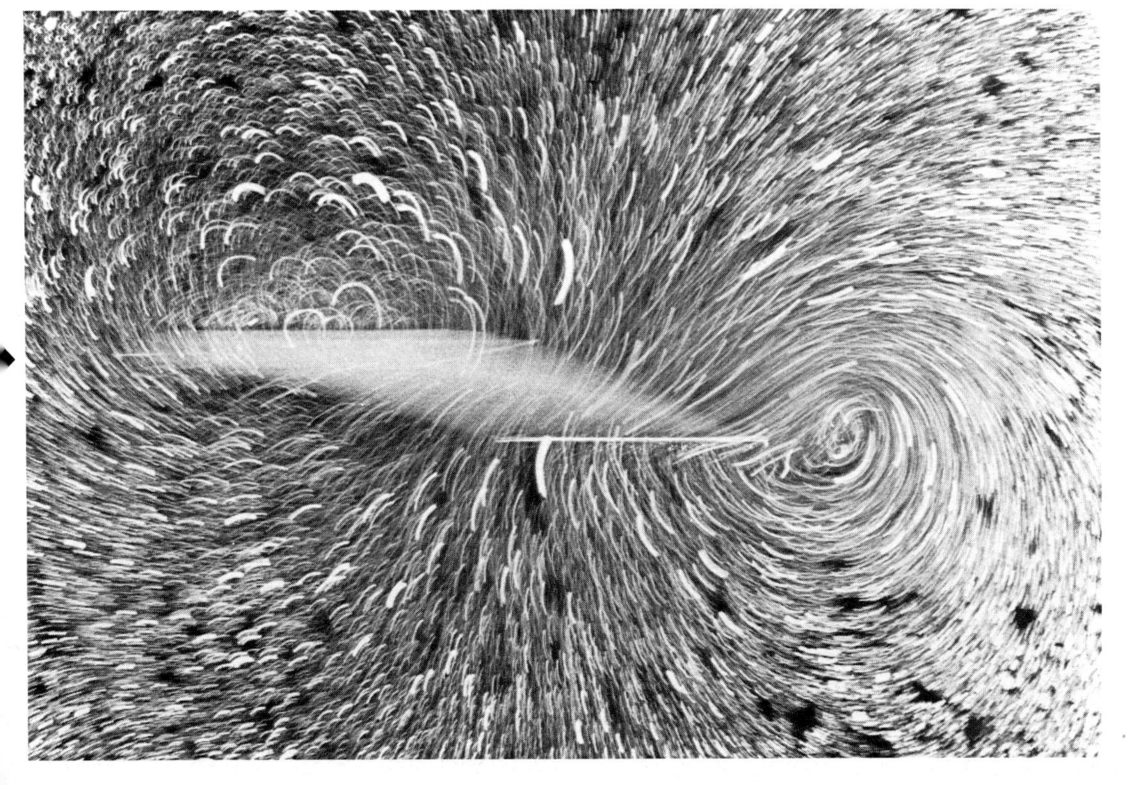

Wings

Lift of an aerofoil

Although the aerodynamic characteristics of aerofoil sections are dependent on the shape of the aerofoil and the Reynolds number, the relationship between angle of attack α and lift coefficient C_L is generally linear at moderate α as expressed by equation (2). The slope of the lift curve, $dC_L/d\alpha$, is 2π rad^{-1} in an ideal fluid, but the experimental value is a little smaller because of the effects of viscosity. Shown in Fig. 122(a) is a typical example of $C_L \sim \alpha$ relation obtained by experiment. C_L is limited by flow separation. C_{Lmax} is due to flow separation developed on the upper surface and C_{Lmin} is due to separation on the lower surface. This phenomenon is called the *stall*. Serial pictures shown in Figs. 122 through 126 show the flow pattern around an aerofoil section, NACA 4412, at various angles of attack. In Figs. 122–124 the flow remains attached on both surfaces of the aerofoil; C_L varies linearly with α. In Fig. 125, $\alpha = 15^{\circ-}$, just before the stall, the flow on the upper surface is just beginning to separate toward the trailing edge, and in Fig. 126, $\alpha = 15^{\circ+}$, just beyond the stall, separation on the upper surface is fully developed.

Types of stall characteristics of aerofoil sections

Near the stall, the relationship between C_L and α of aerofoil sections and their stall characteristics are dependent to the thickness chord ratio, the shape of upper surface near the leading edge, and the Reynolds number. Stall behaviour is generally divided into three types, namely, *thin aerofoil stall*, *leading edge stall* and *trailing edge stall*. Some kinds of aerofoil section change their type according to Reynolds number. Figures 127–130 show the flow around a flat plate of 2% thickness/chord ratio. Because of the sharp leading edge, flow separates from the upper surface at the leading edge at an angle of attack as low as 3°–5° and reattaches further downstream on the surface leaving a *separation bubble*. As the angle of attack increases, the reattachment point moves aft and the bubble grows. At an angle of attack of about 9°, Fig. 129, the separation bubble covers almost the complete chord and C_{Lmax} is reached. This type of stall is called *thin airfoil stall* or *long bubble stall*. The value of C_{Lmax} is small compared with other types of stall, but the change of slope of C_L vs α is gentle, as shown in Fig. 127. Aerofoil sections of thickness/chord ration smaller than about 6% without large camber generally have this type of stall.

Shown in Fig. 131 is a separation bubble on the upper surface of the aerofoil section shown in Figs. 122–126. In this case, at a high angle of attack the laminar boundary layer on the upper surface separates just behind the leading edge and reattaches on the surface, forming a small bubble. As the angle of attack increases a little, the separation bubble shrinks quickly and then bursts, as shown in Fig. 126, resulting in a sudden decrease of C_L. This type of stall is called *leading edge stall* or *short bubble stall*. The lift coefficient increases linearly with α until C_L is very near its maximum value, and the value of C_{Lmax} reached is generally higher than with the other two types of stall. Aerofoil sections of thickness ratio between 10% to 16% have this type of stall and they frequently show a hysteresis in C_L vs α curve at the stall. Figures 132–135 show the trailing edge stall of a *thick aerofoil*, NACA 4421. In this type of stall, flow separation on the upper surface starts at the trailing edge

Figs. 122–126. Flow around an aerofoil section (air, U = 8 m/s, chord length c = 400 mm, $Re = 2.1 \times 10^5$, section, NACA 4412).

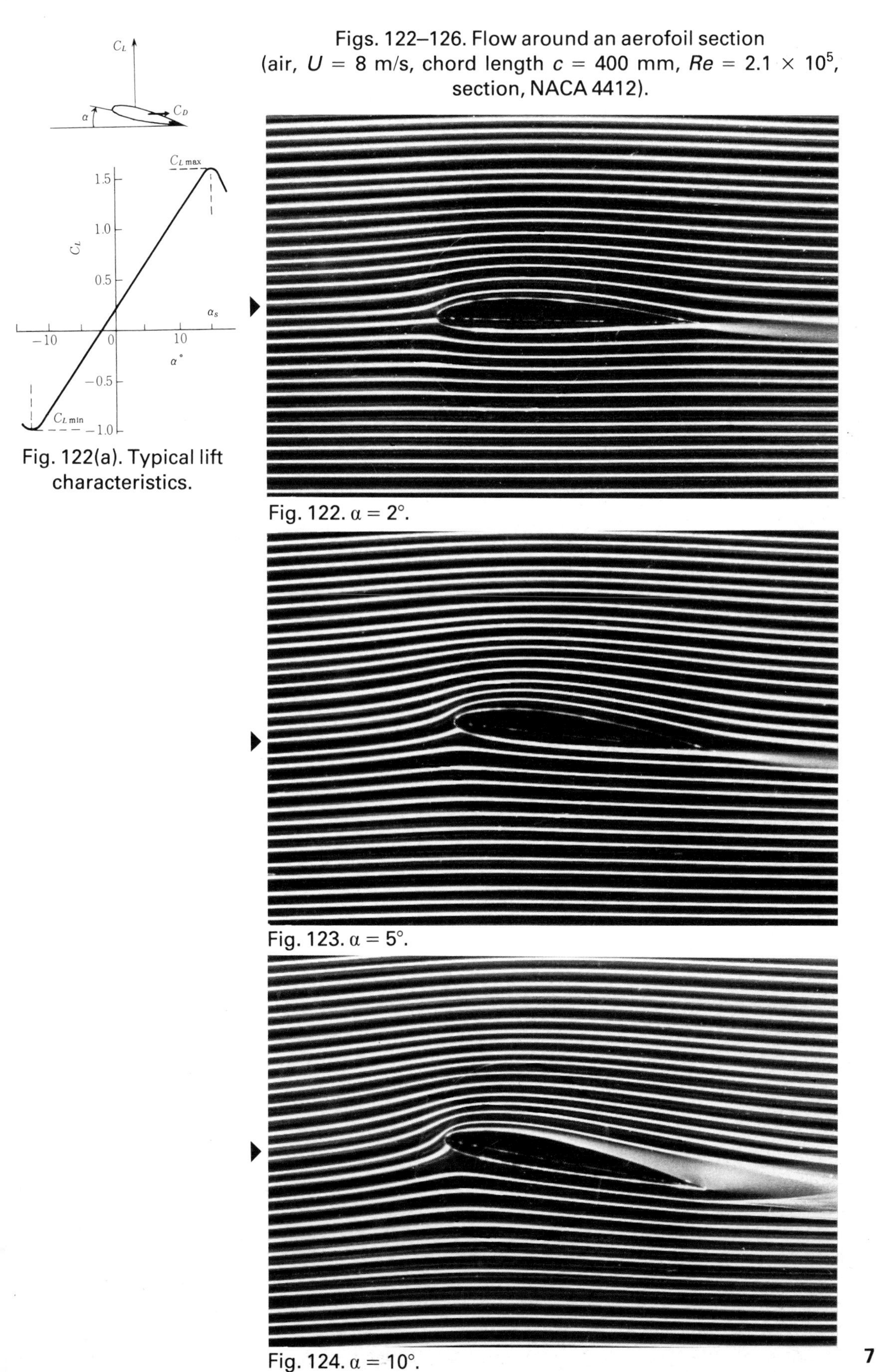

Fig. 122(a). Typical lift characteristics.

Fig. 122. $\alpha = 2°$.

Fig. 123. $\alpha = 5°$.

Fig. 124. $\alpha = 10°$.

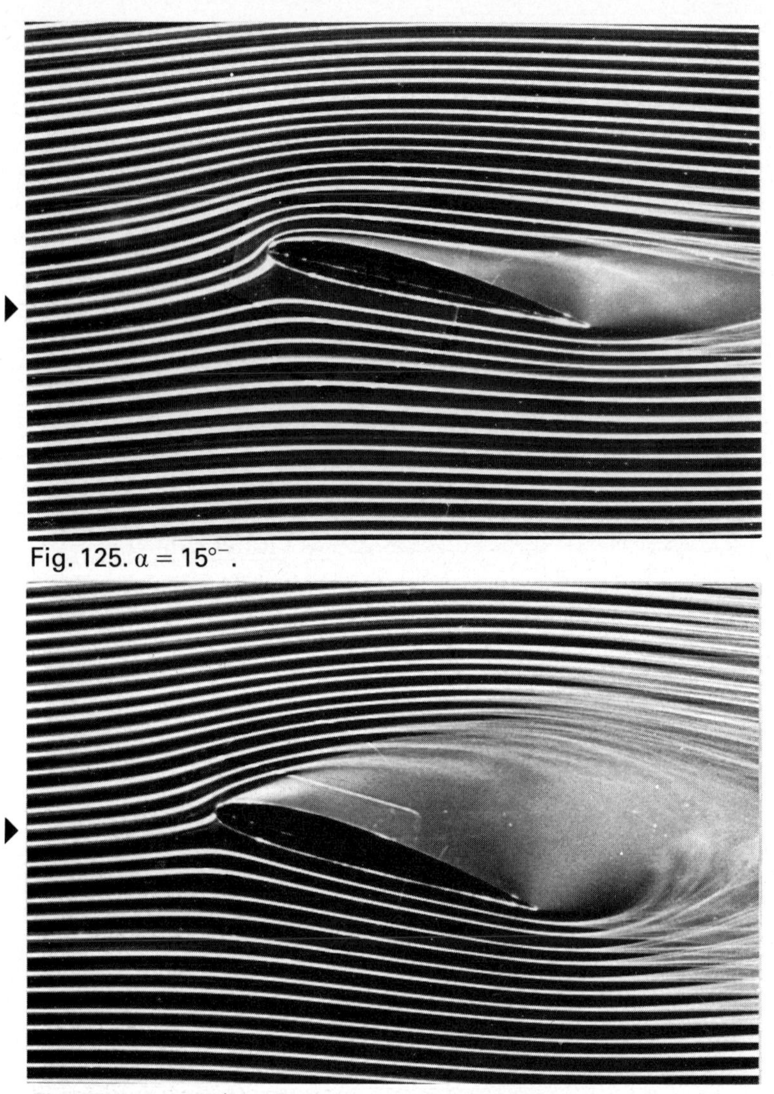

Fig. 125. $\alpha = 15^{\circ-}$.

Fig. 126. $\alpha = 15^{\circ+}$.

and more or less gradually progresses towards the leading edge. The $C_L \sim \alpha$ curve departs from a straight line at a moderate angle of attack, and thereafter changes slope gently with increasing α as shown in Fig. 127.

Functions of flaps

Flaps are movable parts of an airfoil section and are used as *control surfaces* and as *high lift devices*. They are classified as trailing edge devices, leading edge devices, and spoilers.

Trailing edge flaps give an increment or decrement of lift coefficient at a fixed angle of attack, by varying the camber of the aerofoil when they are deflected, as shown in Fig. 136. Figure 136 shows the case of zero incidence (α) and zero angle of deflection (δ). The smoke filament lines are arranged symmetrically and show that $C_L = 0$. Figure 137 shows the case $\alpha = 0$ with $\delta = 15°$. The smoke filament lines are bent upwards and show that $C_L > 0$.

Leading edge devices increase the maximum lift coefficient by postponing the stall to a greater angle of attack as shown in Fig. 138. Deflection of a leading edge flap alleviates the gradient of the pressure rise from the leading edge to the trailing edge on the upper surface. There are many types of leading edge device and some of them are shown in Fig. 138. Shown in Figs. 138–141 are flows around an aerofoil section with a *leading edge slat*. The aerofoil is of NACA

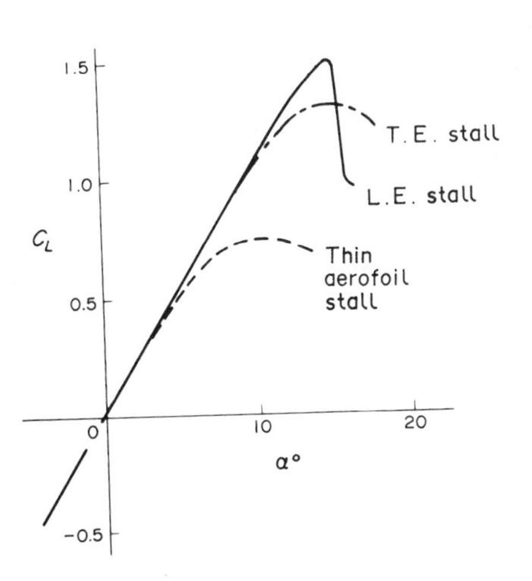

Fig. 127(a). Types of stall and lift characteristics.

Figs. 127–130. Thin aerofoil stall.

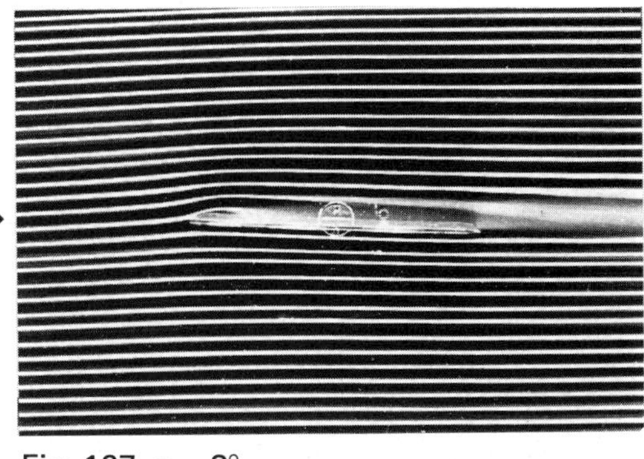

Fig. 127. α = 3°.

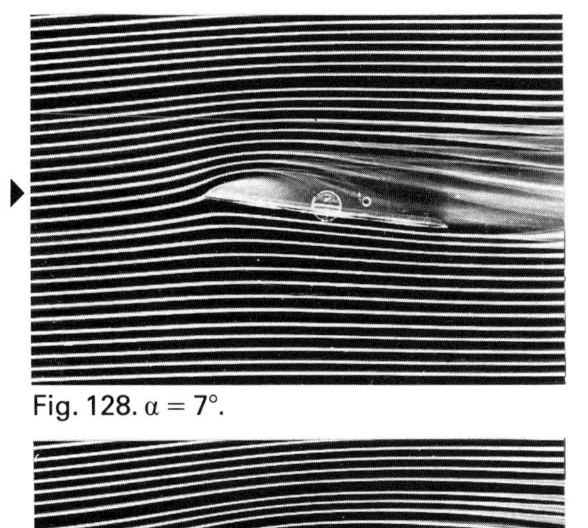

Fig. 128. α = 7°.

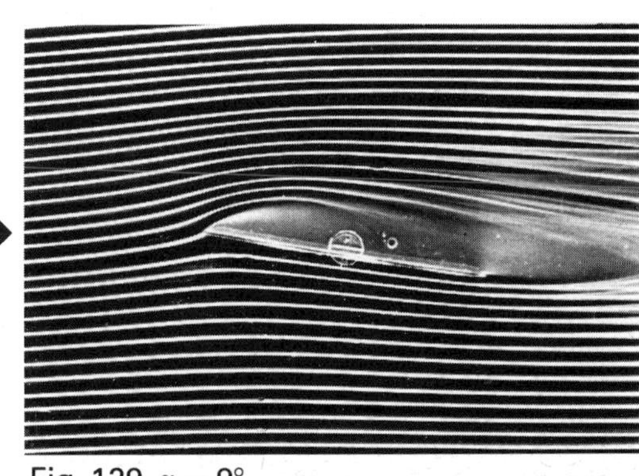

Fig. 129. α = 9°.

Fig. 130. α = 15°.

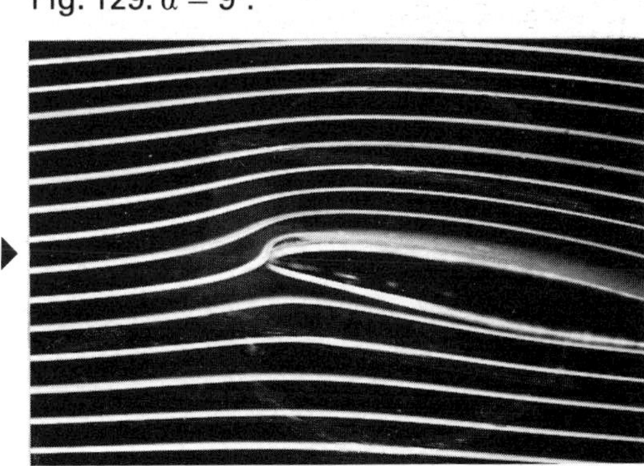

Fig. 131. Leading edge stall (air, $U = 8$ m/s, $c = 400$ mm smoke streak line method).

4412 section (the same as that shown in Figs 122–126) with a leading edge slat having a length of 20% of the chord length. The stalling angle of attack without the slat is about 15°, but with the slat it is postponed to about 30°. The pattern of smoke lines shows clearly the performance of leading edge slat if comparison is made between Figs. 125 and 140 and also Figs. 134 and 140.

Figs. 132–135. Trailing edge stall.

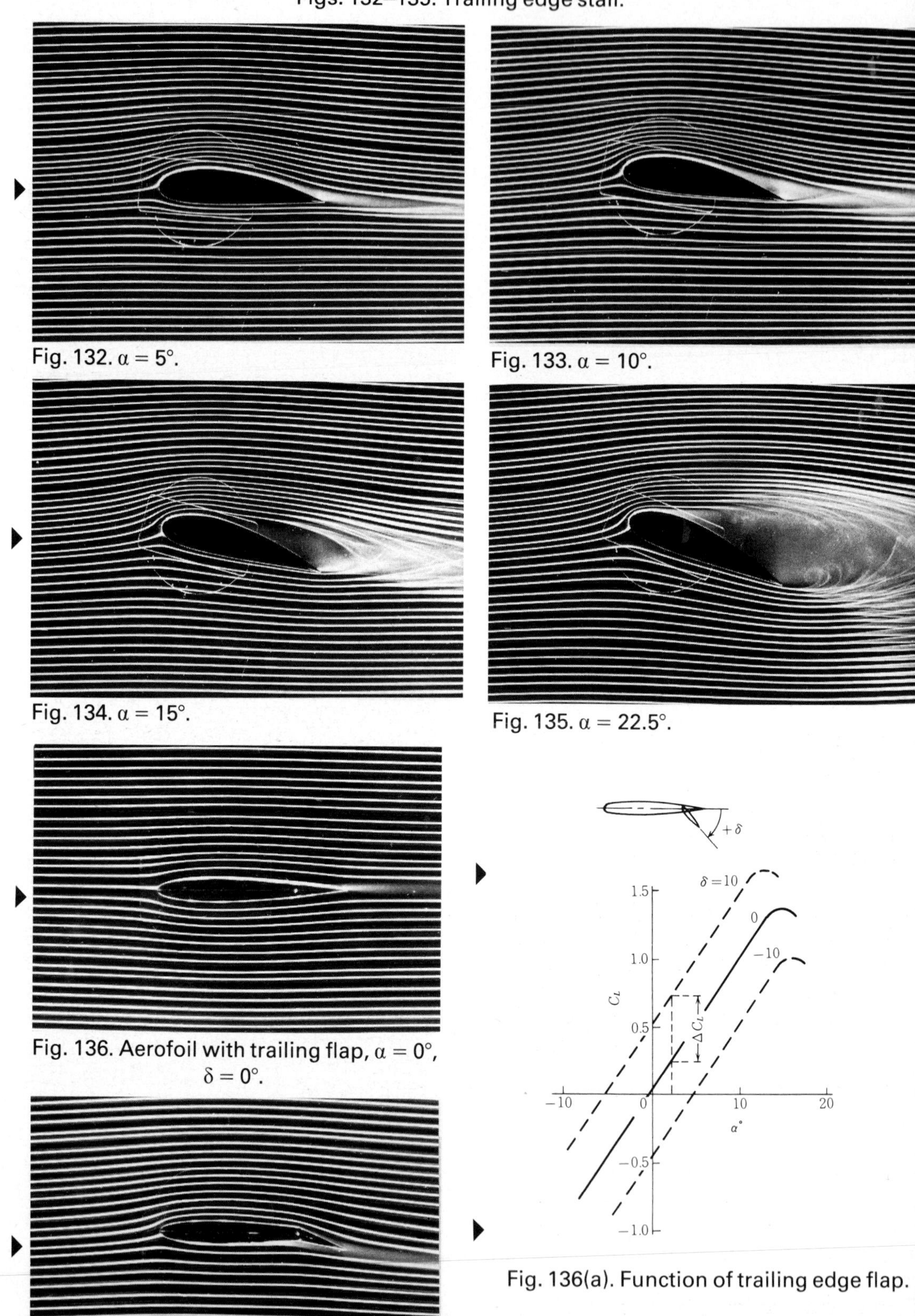

Fig. 132. α = 5°.

Fig. 133. α = 10°.

Fig. 134. α = 15°.

Fig. 135. α = 22.5°.

Fig. 136. Aerofoil with trailing flap, α = 0°, δ = 0°.

Fig. 136(a). Function of trailing edge flap.

Fig. 137. δ, α = 0°, δ = 15°.

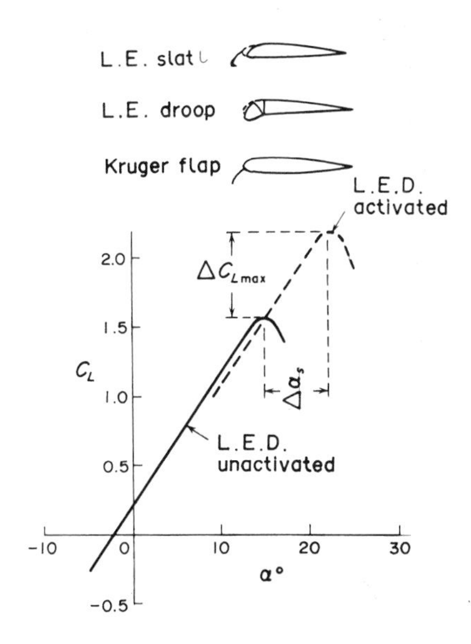

Fig. 138(a). Function of leading edge flap.

Figs. 138–141. Aerofoil with leading edge slat.

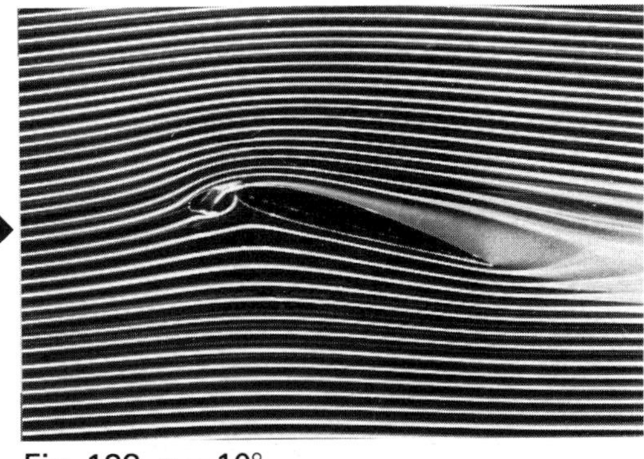

Fig. 138. $\alpha = 10°$.

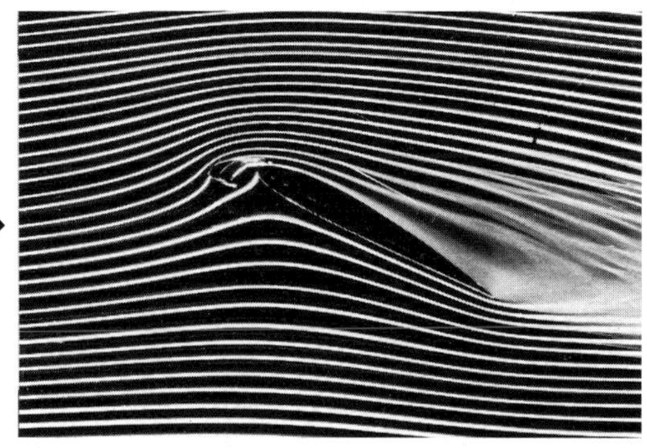

Fig. 139. $\alpha = 25°$.

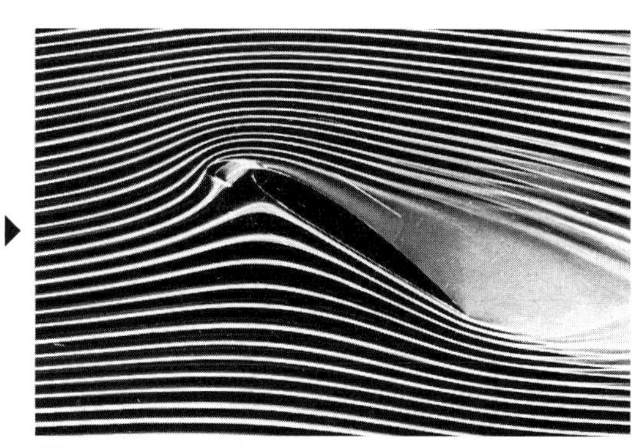

Fig. 140. $\alpha = 30°^{-}$.

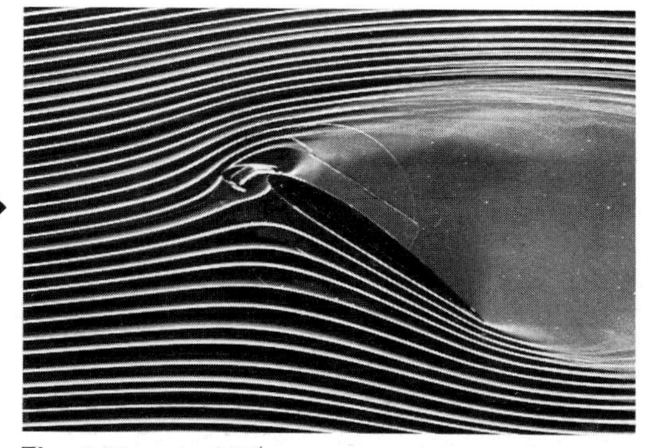

Fig. 141. $\alpha = 30°^{+}$.

Sophisticated mechanical high lift devices

Shown in Figs. 142–144 are the flow patterns around an aerofoil section with sophisticated mechanical high lift devices in landing configuration. Figure 142(a) illustrates the model used. A is the cruising configuration, B is a typical take-off configuration, and C is a landing configuration. In this model, the leading edge slat is of 25% chord length and the trailing edge flaps have a 40% chord length in the housed position. The latter consist of a main flap and three vanes. Figure 143 corresponds roughly to the maximum lift of this configuration and the smoke lines are crowded above the aerofoil showing that a very high lift coefficient is developed. In Fig. 144, the main part of the flow separates from the upper surface of the aerofoil, but the flows coming from the lower side through the slots remain attached on the upper surfaces of their respective vanes and on the main flap. With these devices maximum lift coefficients as high as 4 to 4.5 are obtained.

Figs. 142–144. Aerofoil section with mechanical high lift devices.

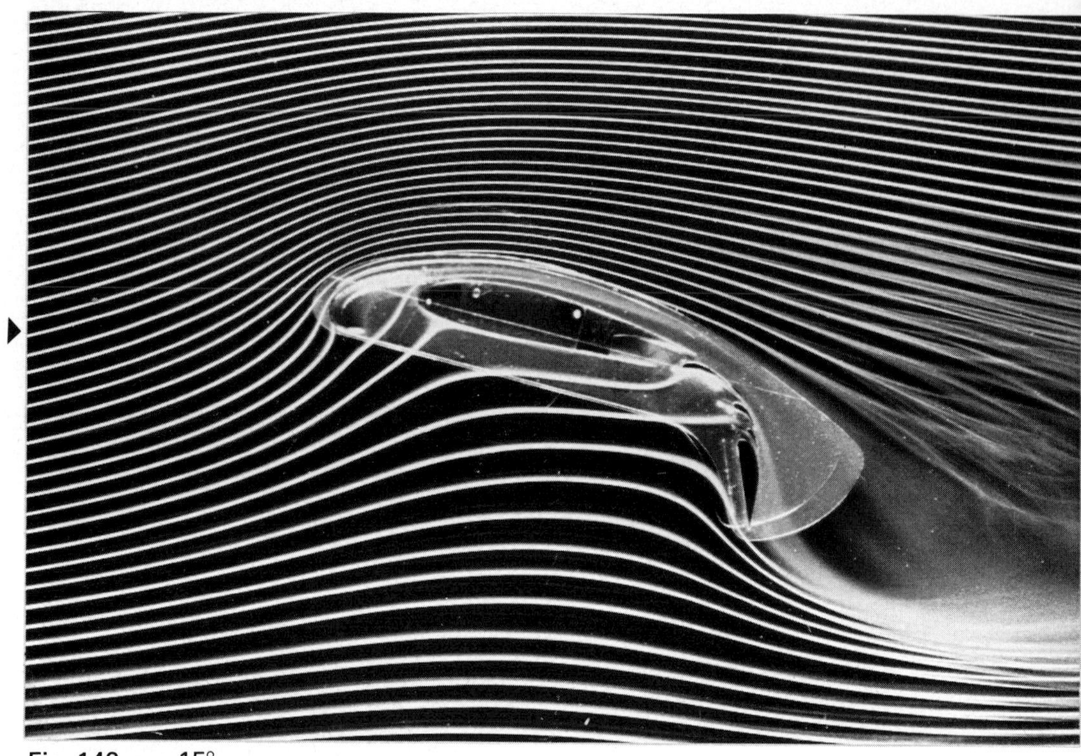

Fig. 142. $\alpha = 15°$.

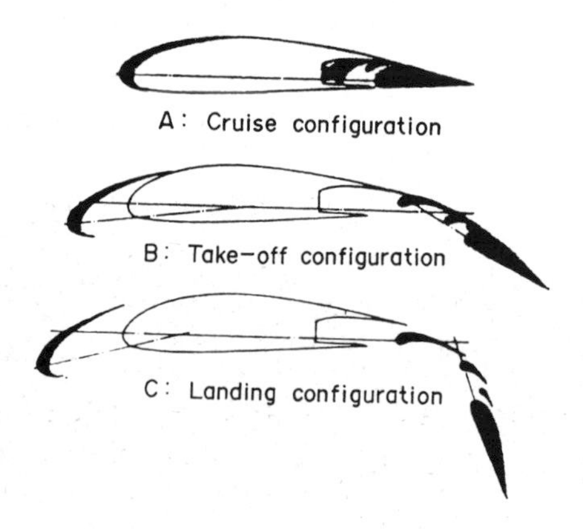

Fig. 142(a). Sophisticated mechanical high lift device.

The development of shock waves with increasing Mach number

Over the upper surface of an aerofoil, the flow is accelerated and the pressure becomes low. This low pressure is responsible for the lift of an aerofoil. When the free stream Mach number is increased to a value approaching 1.0, a local supersonic region followed by a shock wave develops in the region of accelerated flow over the upper surface. The rapid increase of a drag in the *transonic* regime is intimately associated with the appearance of this local, embedded, shock wave. Figures 145–149 show the development of the shock wave over a symmetric aerofoil NACA 0012-65 when the free stream Mach number is increased from 0.6 to 1.2. The aerofoil has an aileron at an angle of two degrees.

At a Mach number of 0.6 (see Fig. 145), no shock waves are observed since the flow is subcritical. The dark region near the leading-edge indicates the high pressure and hence high density near the stagnation point. At a Mach number of 0.8, the flow becomes supercritical, and there appears a weak shock wave that is visible as a black vertical line in Fig. 146. The shock wave-boundary layer interaction makes the boundary layer separate somewhere forward of the shock wave, and an oblique shock wave appears. Thus, the shock wave becomes λ-type. At a Mach number of 0.9, a strong shock wave appears over the lower surface as well as the upper surface. A boundary layer separation which emerges from the foot of the upper surface shock wave can be seen as bright horizontal

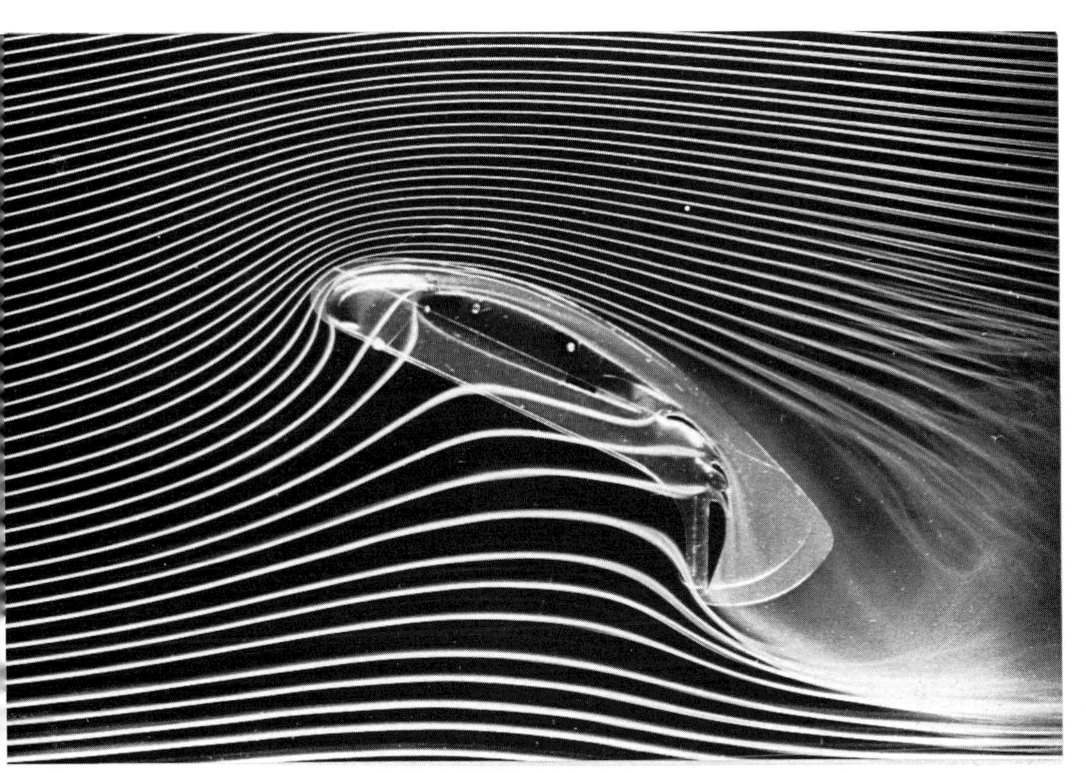

Fig. 143. $\alpha = 20°$.

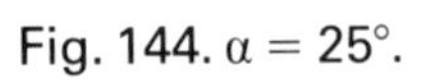

Fig. 144. $\alpha = 25°$.

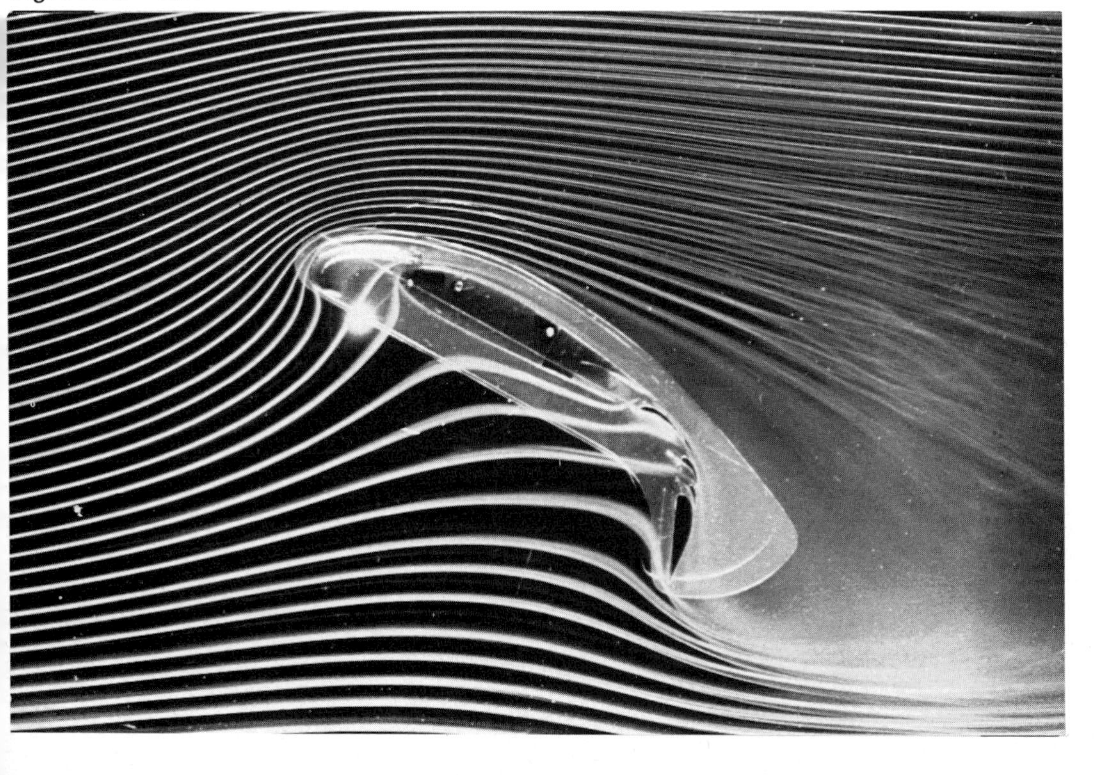

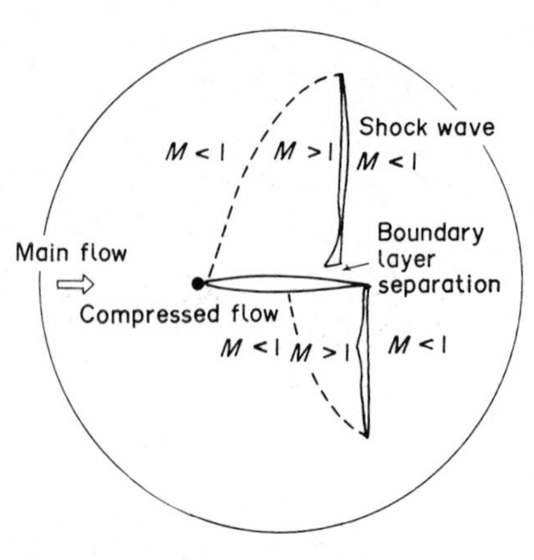

Figs. 145–149. Dry air, free stream Mach number 0.6–1.2, chord length 50 mm, angle of attack 2 degrees, $Re = 6.1–8 \times 10^5$, schlieren photograph.

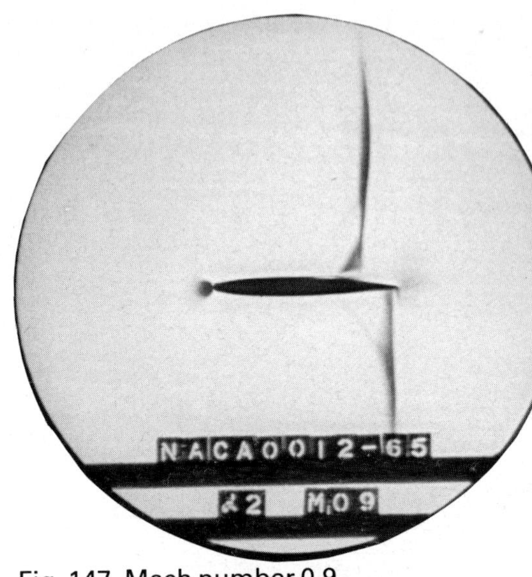

Fig. 147. Mach number 0.9.

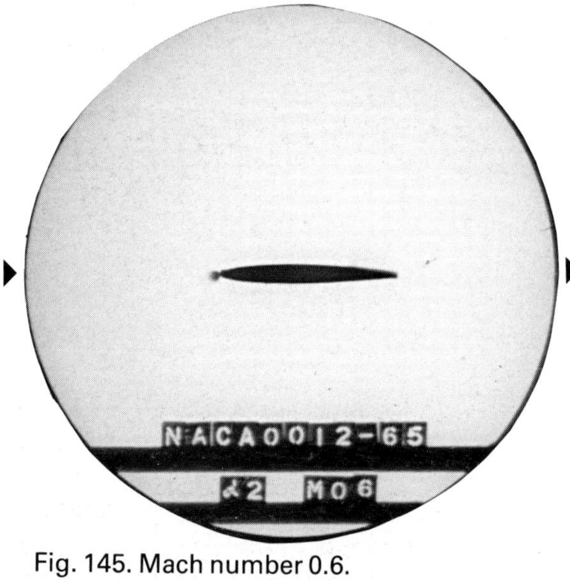

Fig. 145. Mach number 0.6.

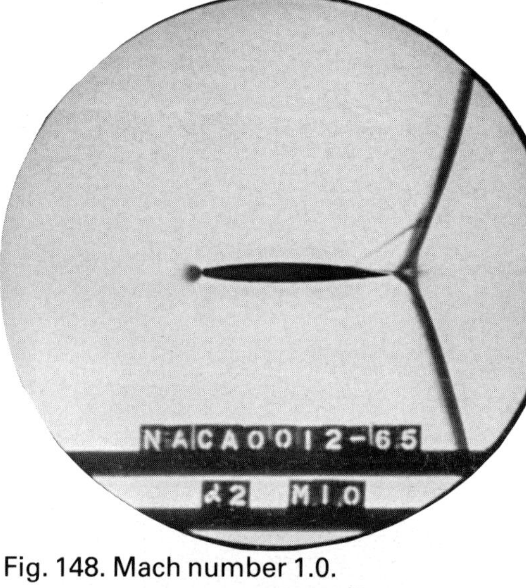

Fig. 148. Mach number 1.0.

Fig. 146. Mach number 0.8.

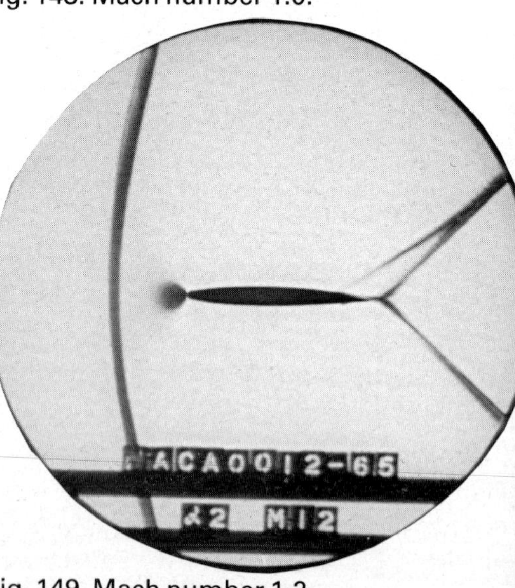

Fig. 149. Mach number 1.2.

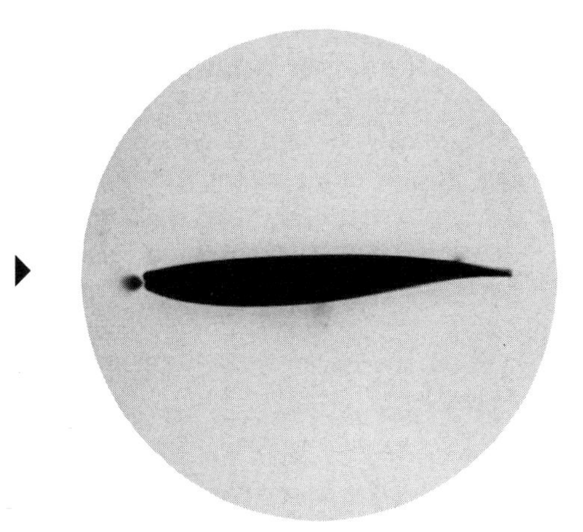

Fig. 150. Angle of attack 0°.

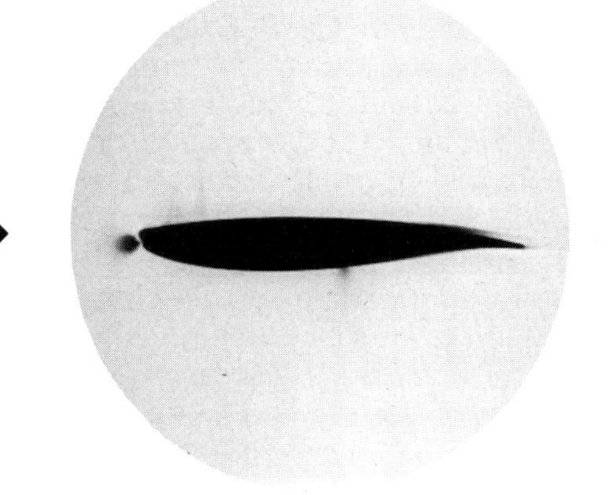

Fig. 151. Angle of attack 2°.

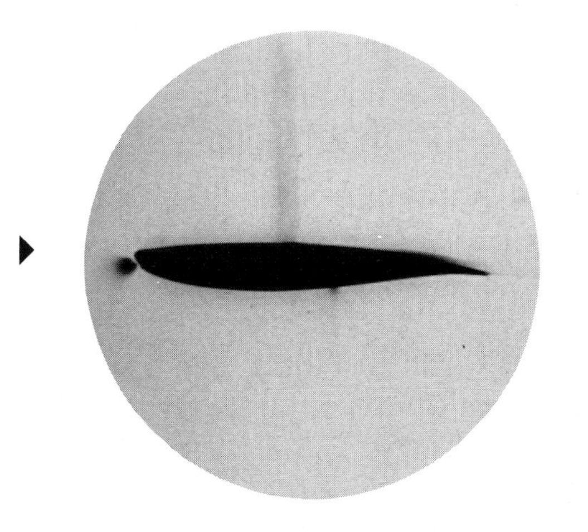

Fig. 152. Angle of attack 4°.

Figs. 150–154. Dry air, free stream Mach number 0.75, chord length 100 mm, angle of attack 0–8°, $Re = 2.3 \times 10^6$, schlieren photograph.

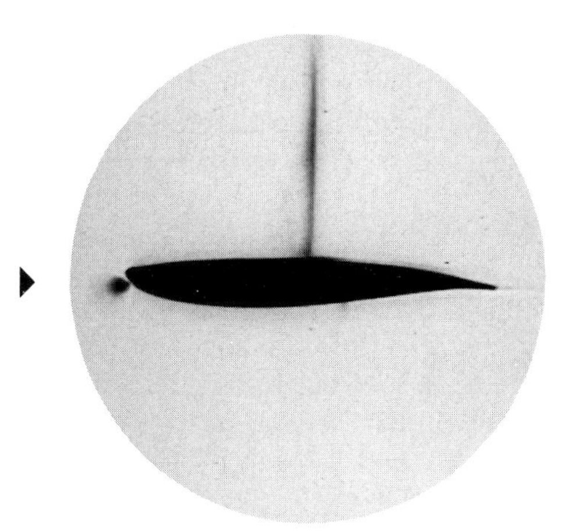

Fig. 153. Angle of attack 6°.

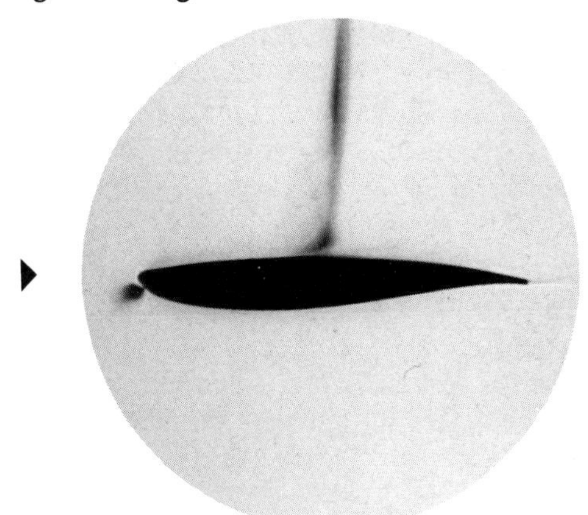

Fig. 154. Angle of attack 8°.

line. The shock wave is located forward of the trailing edge due to the occurrence of this separation. At a Mach number of 1.0, the shock waves are seen to be at the trailing-edge. A weak shock wave and a boundary layer separation are seen at the hinge of the aileron due to the upstream influence of the shock wave at the trailing-edge. At a Mach number of 1.2, a bow shock wave appears in front of the aerofoil since the free stream is supersonic. The flow structure behind this bow shock wave is similar to that at a Mach number of 1.0.

The development of shock waves with increasing angle of attack

When an aerofoil has an angle of attack, the flow is accelerated and the pressure becomes low on the upper surface, whereas, on the lower surface, the flow acceleration is not large and the pressure becomes relatively high. This causes the lift of an airfoil. Figures 150–154 demonstrate the development of the shock waves when the angle of attack is varied at a constant Mach number of 0.75 on a supercritical aerofoil developed by *National Aerospace Laboratory*.

At zero degrees angle of attack, the flow is subcritical over both upper and lower surface, and there are no shock waves. At 2 degrees angle of attack, a shock wave forms at the end of a local supersonic region which appears over the upper surface near the leading-edge. The shock wave is visible as a short faint vertical line near the leading-edge, but it is weak since the supersonic region is small. At 4 degrees angle of attack, the local supersonic region grows larger. As a result, a stronger shock wave appears, and the drag coefficient becomes large. The shock wave is not λ-type since the boundary layer is turbulent and the upstream influence of the shock wave is weak. At 6 degrees angle of attack, the shock wave has become much stronger and is somewhat inclined from the normal to the free stream. The shock wave indicates a turbulent boundary layer separation and near the surface the shock wave becomes oblique. This oblique part of the shock wave is visible as an obscure line at the foot. The coloured picture for the same 6 degrees case is shown on page viii. A yellow region near the leading-edge indicates the compression of the flow, and the green regions above and below indicate the expansion of the flow. A brown region inside the green shows still stronger expansion. A yellow horizontal line emanates from the foot of the shock wave and this corresponds to the separated boundary layer.

At 8 degrees angle of attack, the inclination of the shock wave is quite obvious, the shock wave is very strong and the supersonic region is very large.

Boundary layer flow around an aerofoil

Figure 155 shows velocity distributions within the boundary layer on an aerofoil, visualized by the spark tracing method, using several electrodes. The aerofoil model is made of bakelite. Four conductors pass through the model at the measuring stations and eight line electrodes are set up over the upper and lower surfaces of the aerofoil. The attack angle is smaller than the stall angle (15°), so flow separation occurs only on the upper surface near the trailing edge. The process of boundary layer development on the upper surface can be seen clearly. The velocity distributions for the stalled aerofoil are shown with an angle of attack of 17 degrees in Fig. 156. On the upper surface, the boundary layer has reversed or stagnant

Fig. 155. Velocity distributions within the boundary layer around an aerofoil – attack angle 13° (below the stall)
(air, velocity 20 m/s, chord 150 mm, $Re = 2.1 \times 10^5$, spark tracing method).

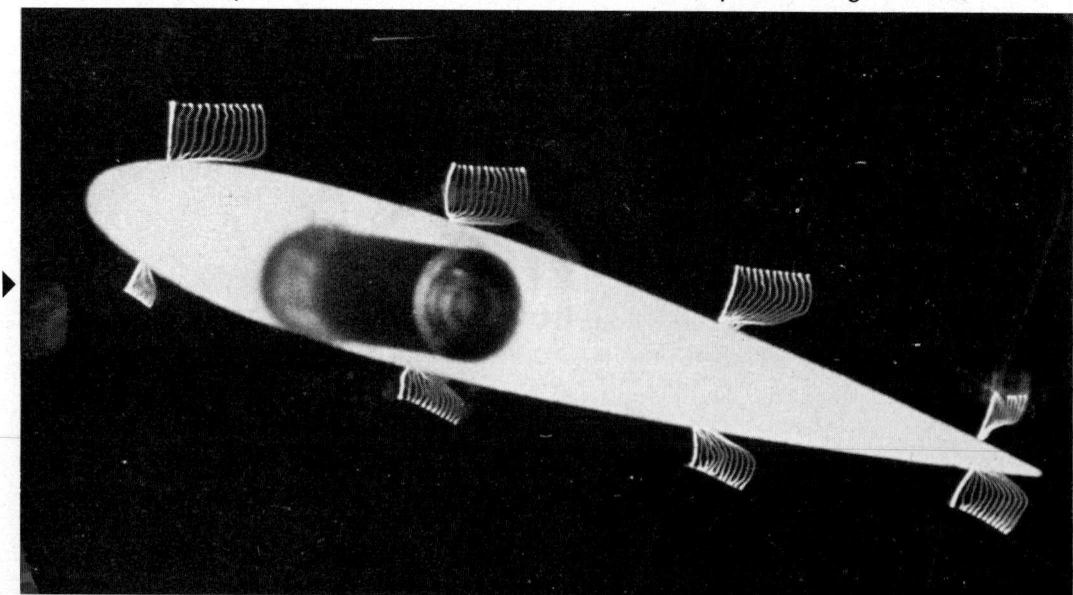

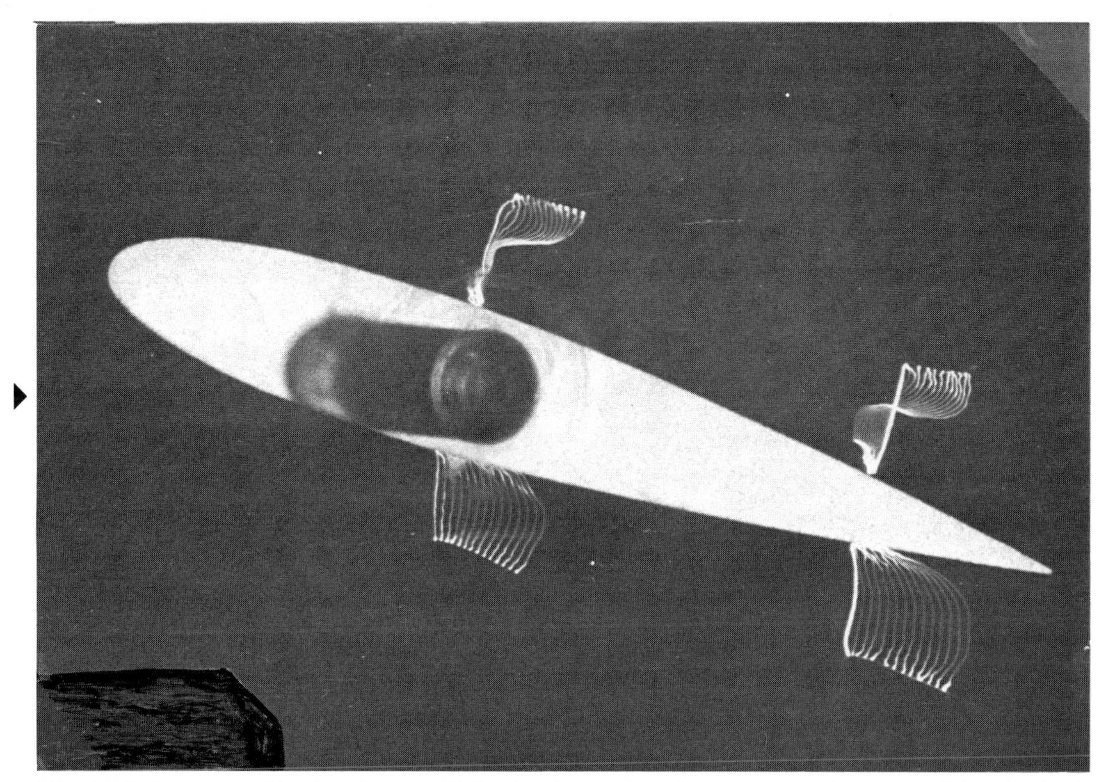

Fig. 156. Velocity distributions within the boundary layer around an aerofoil — attack angle 17° (above the stall).

flow at both measuring stations. The flow patterns on the lower surface, however, are almost the same as in Fig. 155.

Wing tip vortex

On a three-dimensional wing the spanwise distribution of circulation is not uniform, and the difference of circulation between two adjacent sections makes a vortex filament, the axis of which is parallel to the undisturbed flow. The strength of the vortex is highest at the wing tip and zero at the plane of symmetry. These vortices are called *trailing vortices*. The flow around a wing tip is shown in Fig. 157 where a stub wing is mounted horizontally on a side wall of a smoke filament wind-tunnel, and the smoke streak lines are formed in a vertical plane through the wing tip. Several smoke lines are twisted to a strand showing a wing tip vortex.

Flow around an aerofoil

A frontispiece photograph (p. iv) shows an illustration where the flow around a symmetric aerofoil with an attack angle of 10° is visualized using the spark tracing method. The flow separates on the upper surface near the trailing edge and is nearly in the state of stalling. The black part in the wake indicates a region not easily sparked because the vortices occur in it by turbulence and the air resistance increases.

Flow around a delta wing

The flow around a delta wing shows strong three-dimensionality. The behaviour of separated flow is especially different from that in two-dimensional flow. The flow on the upper surface of a delta wing at a high angle of attack is shown in Figs. 158 and 159. The model wing is made of a thin brass plate and has an aspect ratio of 1.5 (the angle of sweep of the leading edge is about 70°). Because of the sharp leading edge, the flow on the upper surface separates at both edges at a relatively small angle of attack, and makes a pair of *lamb-horn vortices*. Energetic outer flow is forced to flow over the upper surface of the central part of the wing by the lamb-horn vortex. Consequently the stall of

Fig. 157. Wing tip vortex
(air, $U = 8$ m/s, c = 200 mm, $Re = 1.1 \times 10^5$, smoke streak line method).

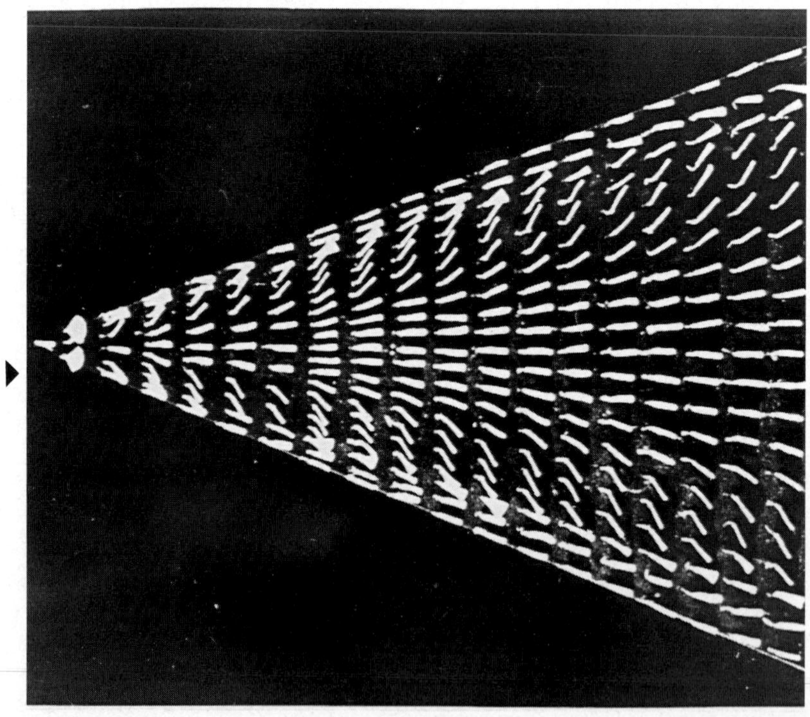

Fig. 158. Flow on the upper surface of a delta wing at a large angle of attack (air, $U = 12$ m/s, $\alpha = 20°$, $Re = 3.1 \times 10^5$, surface tuft method).

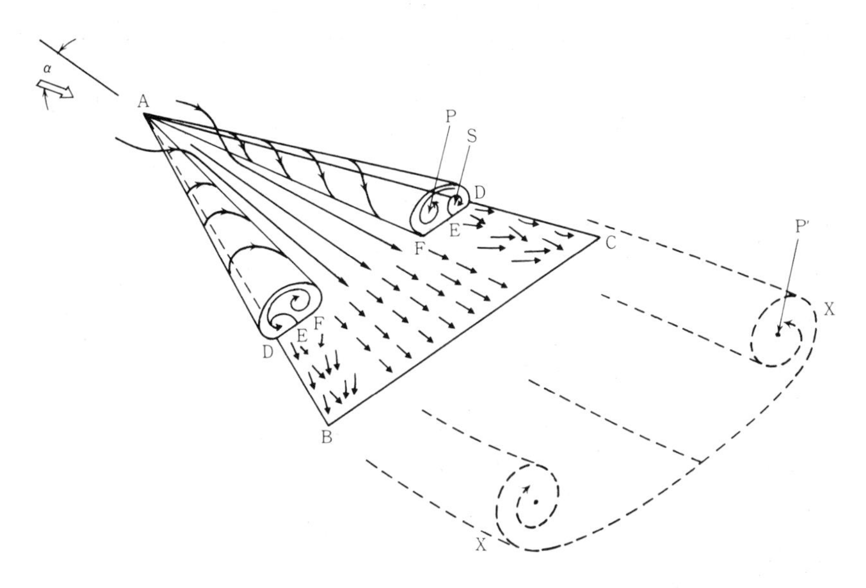

Fig. 158(a). Structure of flow over a delta wing at a large angle of attack.

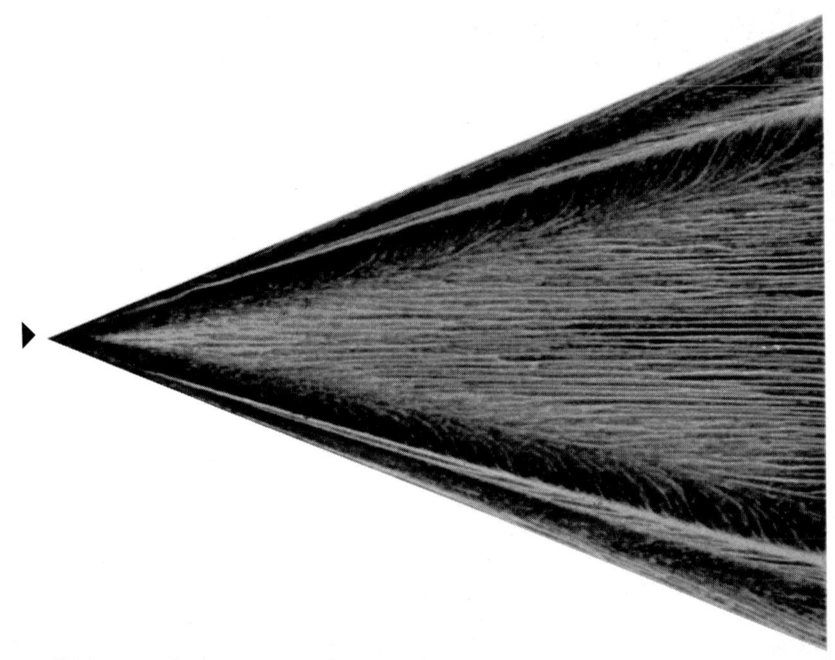

Fig. 159. Structure of flow over a delta wing at a large angle of attack (water, $U = 4.5$ m/s, $Re = 4.5 \times 10^5$, oil film method).

this part of wing is postponed to a very high angle of attack. Line AF in Fig. 158(a) is the attachment line of flow which corresponds to the 8th tuft line from the leading edge in Fig. 158. Line AE in Fig. 158(a) is the line of secondary separation which corresponds to the 3rd tuft line in Fig. 158. The lamb-horn vortex consists of primary (P) and secondary (S) vortices which roll up to form a pair of trailing vortices. On the axis of each primary vortex very low pressure is developed as a result of centrifugal force acting on the rotating fluid. The axial velocity is also very high. The lamb-horn vortex frequently bursts behind the trailing edge or above the wing, resulting in a drastic change of the flow pattern.

Figure 160 shows the flow direction in a plane perpendicular to the undisturbed flow

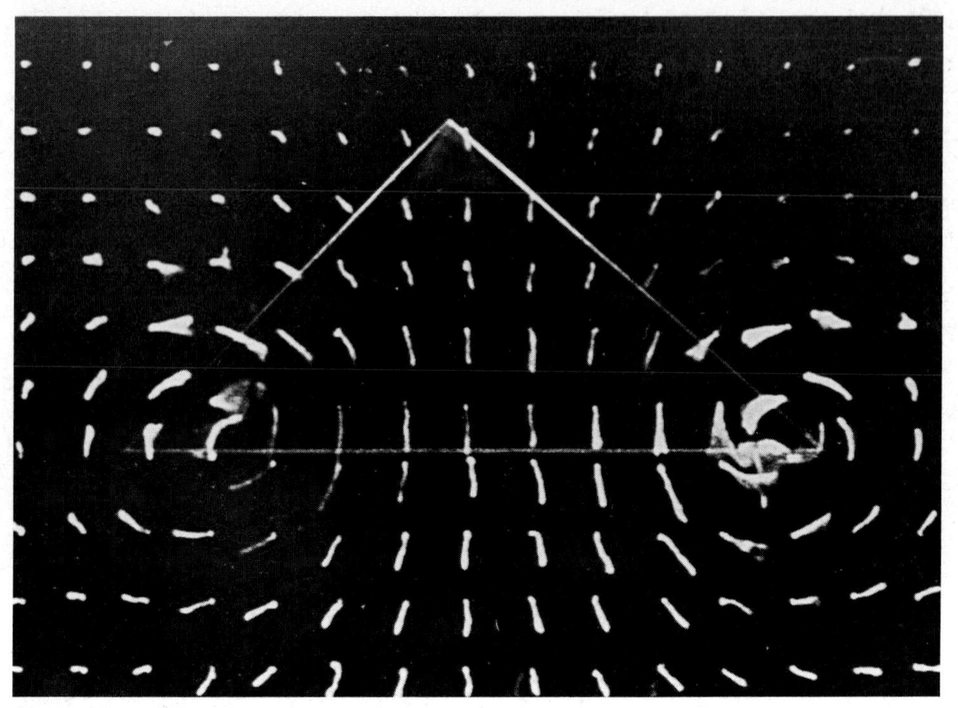

Fig. 160. Trailing vortices behind a delta wing at a large angle of attack (air, $U = 15$ m/s, $\alpha = 20°$, $Re = 3.8 \times 10^5$, tuft grid method).

(Trefftz plane) behind the delta wing. The flow in this plane is governed solely by the pair of trailing vortices. The pattern shown in Fig. 158(a) was deduced from various observations, including those shown above and others obtained by the other methods of flow visualization.

Flow around an oscillating flat plate

Figure 161 shows the streak line pattern around a flat plate performing a rotatory oscillation about the leading edge in a uniform flow. The flow is made visible by means of the electrolytic precipitation method. Immediately after the angle of attack has reached its maximum value, an isolated vortex is shed from the trailing edge of the flat plate. As a result, a vortex street is formed in the wake. When the angular amplitude and the dimensionless oscillation frequency are large, the vortex street moves downstream with a velocity larger than that of the uniform flow. It should be noted that since the flow is unsteady, the streak line pattern does not coincide with the streamline pattern.

Flow around an oscillating aerofoil

The phenomenon of stall, is well known, by which more lift cannot be obtained from an aerofoil due to flow separation. Flow separation also occurs when an aerofoil is in unsteady motion. However, its behaviour in the unsteady case is obscure because sufficient experimental work has not yet been done, though the topic has received much attention in connection with stall flutter; which is one of the self-exciting aero-elastic oscillations of an aerofoil.

Figure 162 shows an example of visualized velocity profiles in water flow around a two-dimensional aerofoil which is executing a pitching oscillation about the 1/4 chord axis. The *hydrogen bubble method* is used in this case, where the rows of minute bubbles of hydrogen are generated by periodic electrolysis from anodes of fine tungsten wire set perpendicularly to the aerofoil chord. By choosing the starting time of the electrolysis, the velocity profiles at any desired part of the pitching cycle can be visualized.

By this method, some of the important features of *unsteady flow separation*, can be seen, such as the extent of the reversed flow

Fig. 161. Water, flow velocity 1 cm/s, plate length 5 cm, Reynolds number 500, dimensionless frequency fd/U 0.51, angular amplitude 30°, electrolytic precipitation method.

Fig. 162. Flow around an oscillating aerofoil
[water, fluid velocity 1 cm/s, chord length of aerofoil 10 cm, $Re = 1 \times 10^3$, reduced frequency 0.18, mean angle of attack 10°, amplitude 5°, present angle of attack 13.6° (decreasing), hydrogen bubble method].

region behind the aerofoil, as well as its time history.

Oil-flow pattern on the suction surface of a sweptback wing

On the suction surface of a sweptback wing, a secondary flow in the boundary layer towards the root in the leading edge region (a) and towards the tip downstream of the region of minimum pressure (b) is induced by the spanwise pressure gradient, as indicated in the illustration below. The boundary layer thickness at the trailing edge is increased markedly toward the tip. At an angle of attack near the stall the boundary layer flow near the trailing edge becomes almost parallel to the trailing edge, and starts to separate at the tip with the further increase of attack angle. Figure 163 shows an aspect of the boundary layer flow described above, visualized by using the oil-film method. The white ribbon-like deposit of oil film near the leading edge indicates the existence of a laminar separation bubble.

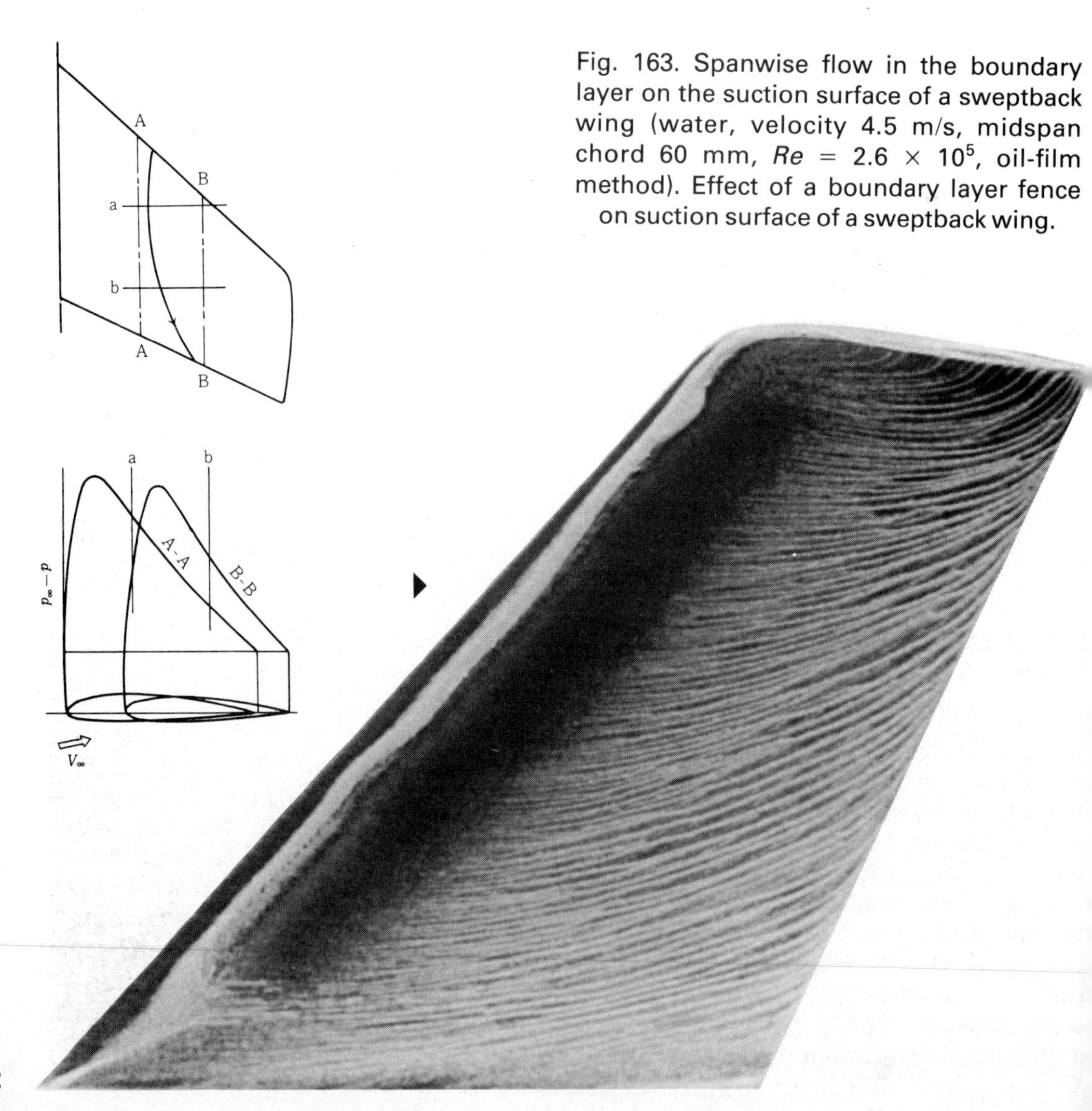

Fig. 163. Spanwise flow in the boundary layer on the suction surface of a sweptback wing (water, velocity 4.5 m/s, midspan chord 60 mm, $Re = 2.6 \times 10^5$, oil-film method). Effect of a boundary layer fence on suction surface of a sweptback wing.

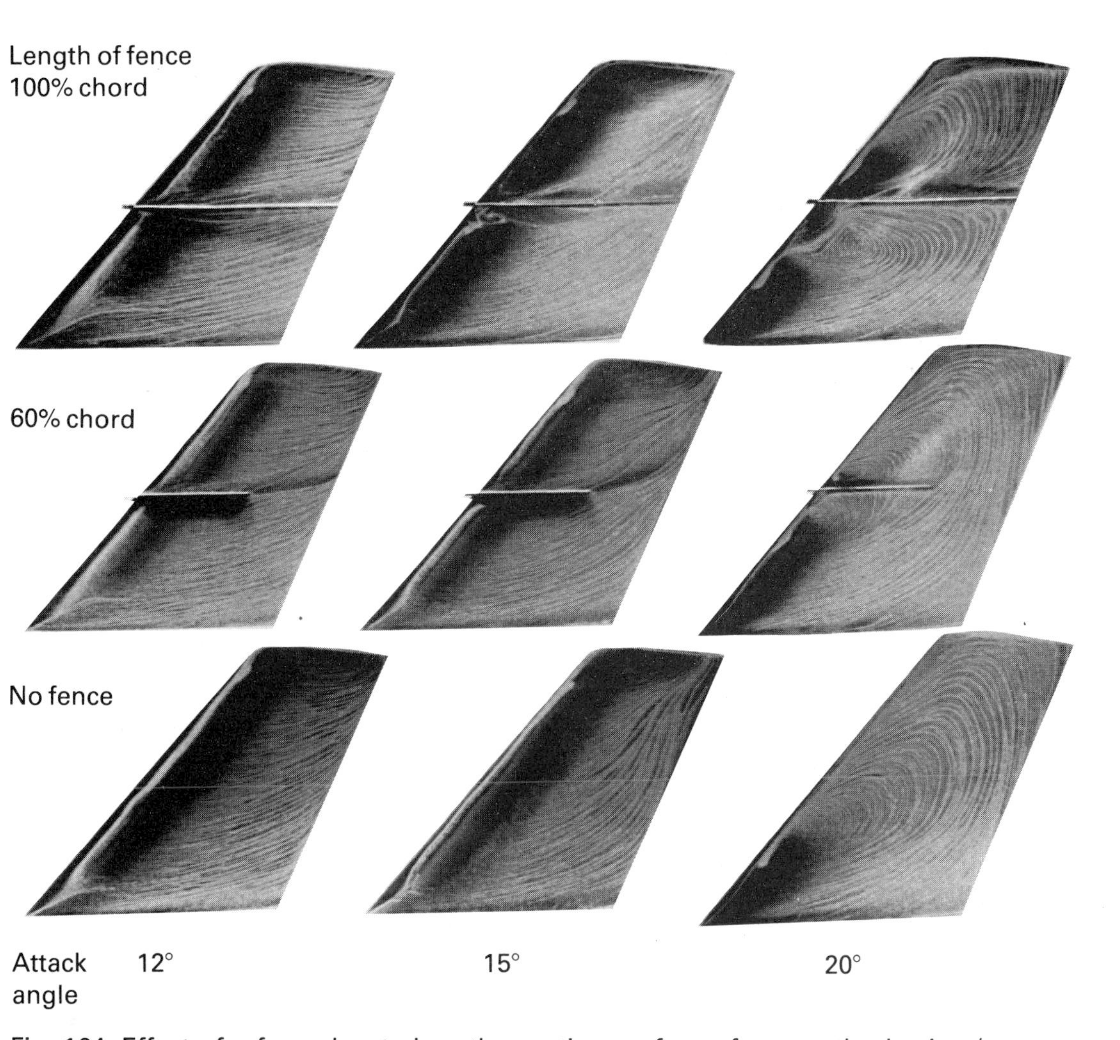

Fig. 164. Effect of a fence located on the suction surface of a sweptback wing (water, velocity 4.5 m/s, midspan chord 60 mm, $Re = 2.6 \times 10^5$, oil film method).

The effect of a fence located on the upper surface of a sweptback wing

Figure 164 shows examples of oil film patterns on the upper surface of the plain swept wing shown in Fig. 163, together with two similar wings, equipped with boundary layer fences having lengths of 60% and 100% chord, at three different angles of attack.

At 15° angle of attack, without the fence, the movement of the boundary layer towards the tip is so strong that the surface flow direction becomes parallel to the trailing edge, and near the tip the pattern has the appearance association with separation. With the fence added, the spanwise movement of the boundary layer is reduced, but the difference between the effect produced by the two different lengths of fence is small up to an angle of attack of 15°. The difference becomes more marked at higher angles. At 20° where, on the plain wing, except near the wing root, the boundary layer flow turns back toward the leading edge, the 60% chord fence can only slightly increase the unseparated region near the root. At this angle of attack, even the 100% chord fence does not suppress separation, but it divides the boundary layer flow into two similar reversed flow patterns, without, however significantly improving the overall wing behaviour.

Cascades and Fluid Dynamic Machinery

Adiabatic wall temperature distributions of a film cooled flat plate

Film cooling is used in many systems to protect solid surfaces exposed to high temperature gas streams. Coolant injected into the boundary layer acts as a heat sink, reducing the gas temperature near the surface. Applications are widespread, particularly in *gas turbines* where combustion chamber liners, turbine blades and other hot parts require cooling.

Figure 165 shows two dimensional temperature distributions on a flat plate, in the region around injection holes and in the region downstream of them. The original figure was obtained with an *infrared scanner* as shown in the diagram. Image processing was then applied to obtain a colour/grey zone display consisting of twelve levels, with smoothed contours showing the boundary of each level.

Material of low thermal conductivity is used for the test model in order to observe only the film cooling effect, minimizing heat conduction and convection effects behind the plate surface. Appropriate film cooling hole arrangements and blowing rates can effectively be sought with this measurement technique.

For ease and accuracy of testing, heated coolant and an unheated mainstream are often used, since the temperature field is similar. Figures 165 and 166 were obtained in this way.

Adiabatic wall temperature distributions of a film cooled gas turbine vane model

Figure 166 shows temperature distribution on a gas turbine vane acrylic model. Two kinds of *liquid crystal*, sensitive to different temperature ranges (left half: 35°–39°, right half: 42°–45.5° [C]), were used at the same time to cover a larger range than would otherwise be possible. Temperature and flow conditions were carefully adjusted so that the observable part of the distributions would appear around the *film cooling* injection holes and the region downstream.

Cascade tests with enlarged vane models, and under moderate pressure and temperature conditions, provide spatially detailed data on film cooling effectiveness, from which the effects of surface curvature and static pressure gradient can be examined with main flow patterns similar to those in the actual gas turbine vanes.

The results obtained indicate the attachment of the injected coolant to the vane's surface to a point far downstream, a distance of 9–10 times of the hole diameter. (In the original this was indicated by blue colour, which implies high temperature.) Further downstream, the interaction between the mainstream and the coolant flow becomes dominant, and eventually the surface temperature approaches the mainstream temperature (originally shown with graded colour changes, green-yellow-red-black).

Flow in the passage between the vanes of a centrifugal pump impeller

A centrifugal pump comprises an impeller rotating in a *volute chamber*. The impeller has several curved vanes which are supported between two shrouds. Liquid is led into the eye of the impeller, and moves away from the centre through the intervane passages

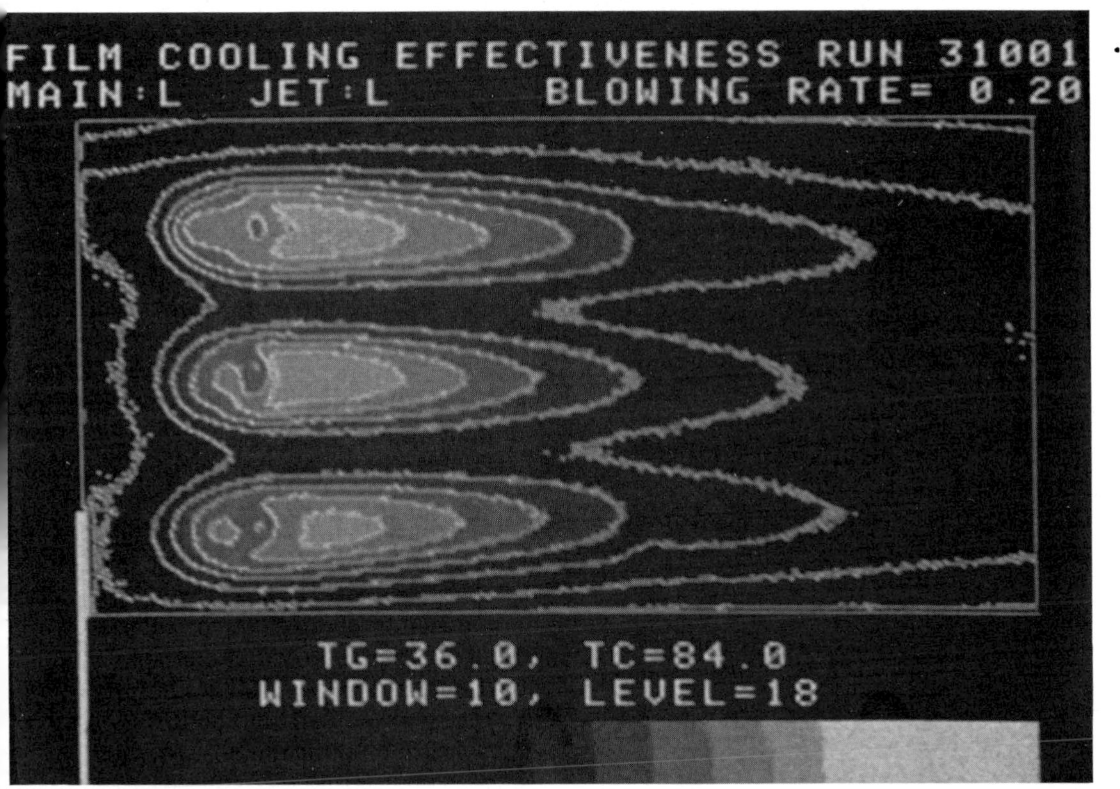

Fig. 165. Air, velocity 4.5 m/s (mainstream), 0.9 m/s (injected flow), temperature 36°C (mainstream), 84°C (injected flow), flat plate with injection holes 12 mm (inner diameter), 36 mm (pitch), 35° (injection angle), $Re = 3.4 \times 10^3$, infrared scanner.

increasing its velocity and pressure head as a result of the rotation of the impeller. Then, centrifugal force discharges the liquid into the volute chamber, in which a part of the kinetic energy is converted into pressure head.

The flow is observed through the transparent plastic rear shroud and a transparent casing rear cover plate, and is made visible by white and red surface tufts attached, respectively, to the front and rear shroud surfaces. The right-hand intervane passage has only the red tufts on its rear shroud face, and the left-hand intervane passage has only white tufts on its front shroud face. The centre intervane passage carries both sets of tufts.

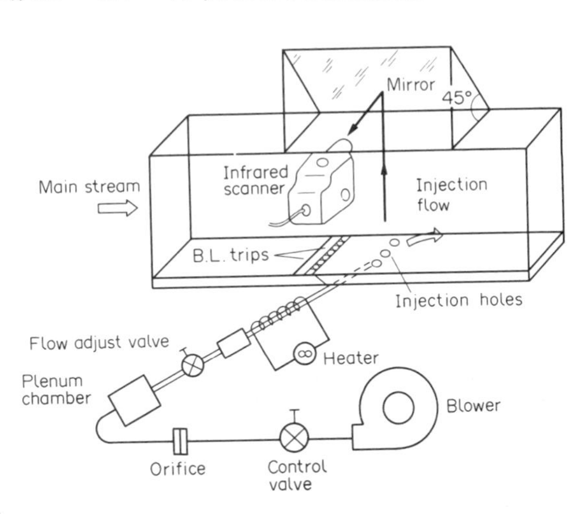

Figure 167 shows the situation when the flow rate through the pump is large. The flow indicated by the front shroud tufts, in the left hand passage can be seen to follow the contours of the vanes smoothly, though velocity is high near the concave face of the leading vane, and low near the convex face of the trailing vane. Figure 168 shows the situation with a small flow rate. A distinct reversed flow region can be seen on the leading, convex, face of the trailing vane, toward its inner end. At this flow rate also, many differences may be observed between the flows on the front and rear shrouds. These are due to the marked difference in entry conditions at the front and rear of each passage.

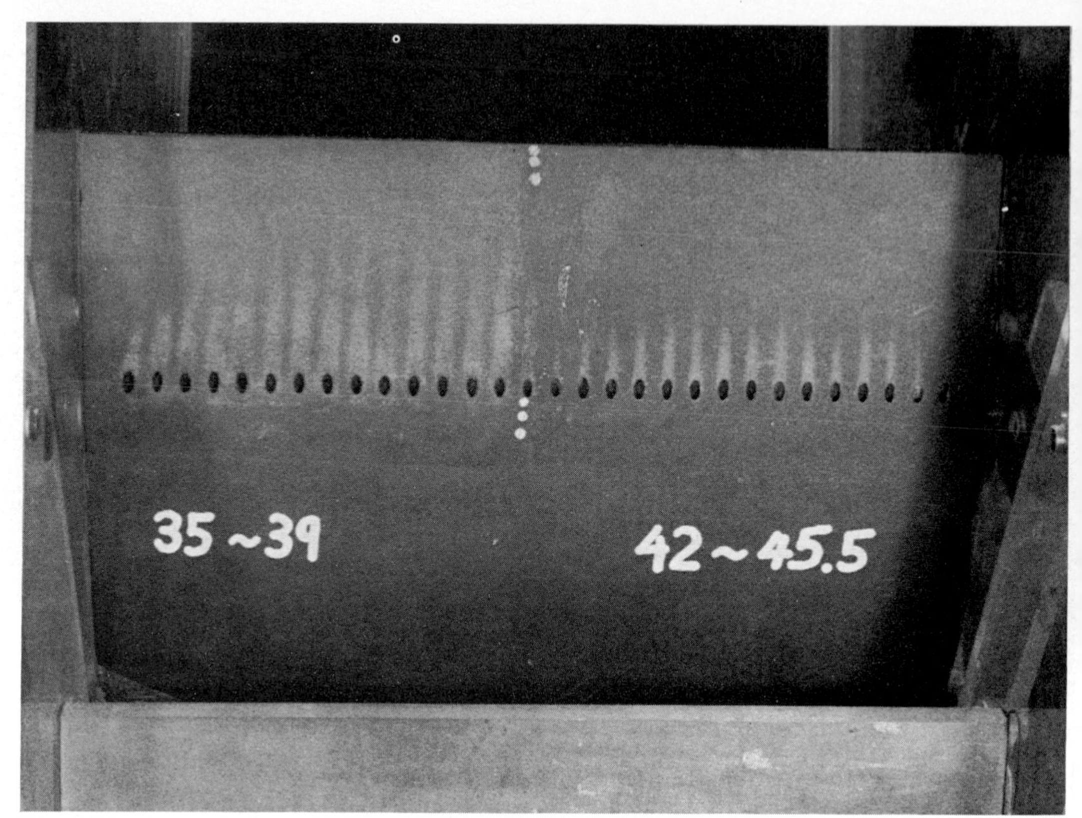

Fig. 166. Air, velocity 35 m/s (mainstream), 32 m/s (injected flow), temperature 24°C (mainstream), 55°C (injected flow), test vane 218 mm (chord), 270 mm (span), 2.8 mm (inner diameter) 30° (injection angle), $Re = 5 \times 10^5$, liquid crystal.

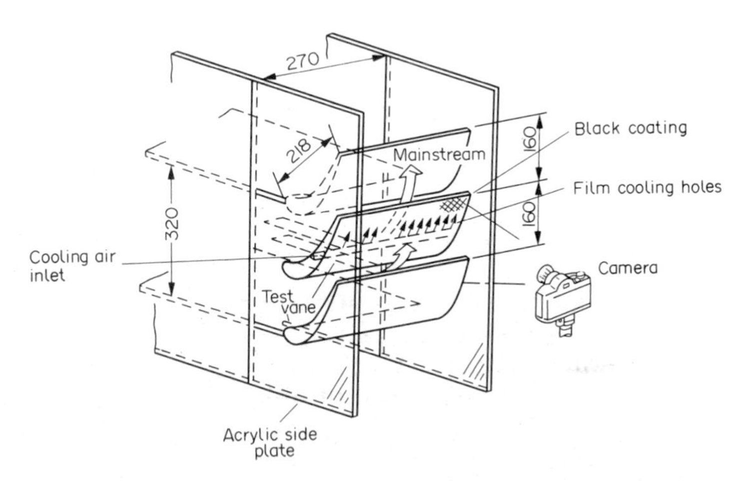

Flow in the volute casing of a centrifugal pump impeller

The discharge flow from a centrifugal impeller is axisymmetric and shows a free-vortex flow pattern only at the designed capacity (Q_0) of the volute. In this case the stream line of the discharge flow becomes a logarithmic spiral, the angle of which coincides with the volute angle. In the *partial discharge* range, the local discharge capacity from the impeller decreases monotonically toward the discharge nozzle, while it increases in an *over-discharge* range. Accordingly, the flow patterns in the volute and around the volute tongue show completely different behaviour in the partial and over capacity ranges. This is clearly illustrated by visualizing the impeller outlet flow.

Fig. 167. In the case of large flow rate.

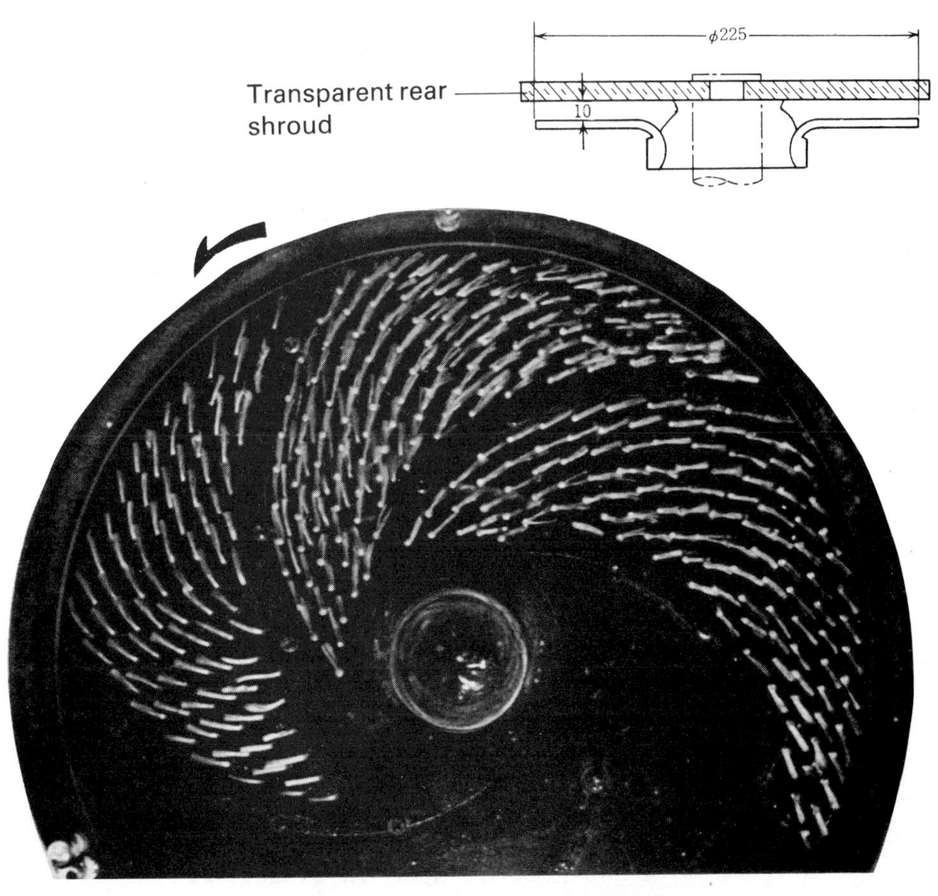

Fig. 168. In the case of small flow rate
(water, rotational speed 1200 rpm, impeller diameter 225 mm, tuft method).

Figs. 169–171. Water, impeller peripheral velocity 10.2 m/s, mean flow velocity in volute 3.3–3.7 m/s, impeller diameter 260 mm, exit depth 16 mm, $Re = 1.1 \times 10^6$, oil-film method.

Fig. 169. In partial flow rate ($Q = 0.6Q_0$).

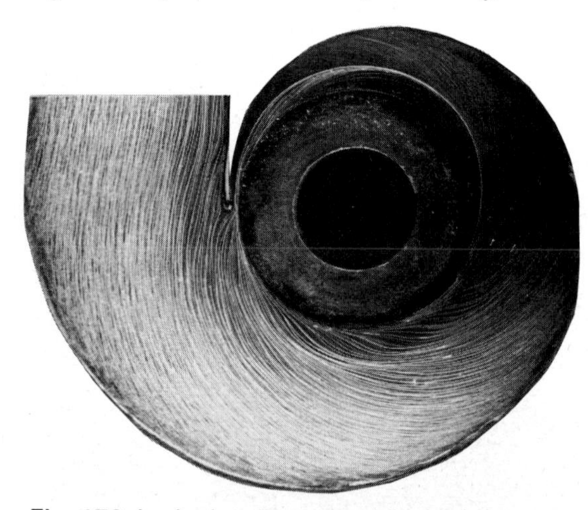

Fig. 170. In design flow rate ($Q = Q_0$).

Fig. 171. In over flow rate ($Q = 1.6Q_0$).

Figures 169, 170 and 171 show stream lines visualized by the *oil-film* method on the volute wall of a two-dimensional logarithmic spiral, the width of which is equal to the impeller exit width. The concentric circles in the figure correspond to the impeller inlet and outlet boundaries.

At partial discharge ($0.6Q_0$), the stream angle around the impeller periphery just outside the impeller exit becomes smaller towards the discharge nozzle and a three-dimensional separation is indicated, which results in a large recirculating flow through the tip gap of the volute tongue. Conversely, in the over-discharge state, the stream angle around the impeller periphery just outside the impeller exit becomes larger towards the discharge nozzle and a large flow separation is indicated at the discharge side of the volute tongue. In the region at large radius, the stream lines are seen to be inclined inward, due to the secondary flow near the wall.

Flow through a turbine cascade (changes from subsonic to supersonic state)

A series of figures (Figs. 172–178) show how the flow through a turbine cascade changes from fully subsonic to supersonic.

Figures 172–175 give the equi-density contours of the flow visualized as fringes by a *Mach–Zehnder* interferometer. Flow enters the cascade, from the lower left and leaves to the right.

Figure 172 gives the fully subsonic case at an outlet *Mach number* of 0.68, showing that the flow is smoothly accelerated with no *separation* and that the wake is indicated by fringes nearly parallel to the flow.

Figure 173 gives the case of a higher outlet Mach number of 0.85, showing the larger acceleration of the flow with the more closely spaces equi-density contours. For an outlet Mach number higher than 0.85, a supersonic flow region appears on part of the suction surface of the blade.

Figure 174 gives the case of an outlet Mach number of 1.05, when the flow is supersonic on the downstream half of the suction surface, and a *shock* is visible near the trailing edge, as fringes standing perpendicular to the flow.

For a still higher outlet Mach number of 1.54, in Fig. 175, the shock generated at the

Fig. 172. Subsonic flow.

Fig. 173. High subsonic flow.

Air, outlet flow velocity (172) 233 m/s, (173) 292 m/s, 174 (360 m/s), (175) 528 m/s, blade chord 33.6 mm, $Re = 5.0 \times 10^5$, Mach–Zehnder interferometry.

trailing edge of the upper blade becomes slanted, extending downstream of the trailing edge of the lower blade.

The flow shown in Fig. 60 corresponds to that of the intermediate state between Figs. 174 and 175, when the shock from the upper blade is incident on, and is reflected from the lower blade.

Figures 176–178 show pictures taken by the *schlieren method*, which is particularly suitable for the observation of shock waves in a *compressible flow*.

Figure 176 corresponds to Fig. 172 of fully subsonic flow, in which the wake extending downstream from the trailing edge is clearly visible.

In the case of the higher Mach number of 1.29 in Fig. 177, the shock generated on the

Fig. 174. Transonic flow.

Fig. 175. Supersonic flow.

upper blade impinges on the lower blade, so that the abrupt pressure rise leads to separation of the boundary layer on the lower blade to form a separation 'bubble' with two reflected shocks, as illustrated by the accompanying sketch.

Figure 178 corresponds to Fig. 175, showing clearly the features of supersonic exit flow, that is, the compression shock and expansion fan emanating from the trailing edge of each blade, with no interaction on the blade below, together with the deflection of the wake as it crosses a shock.

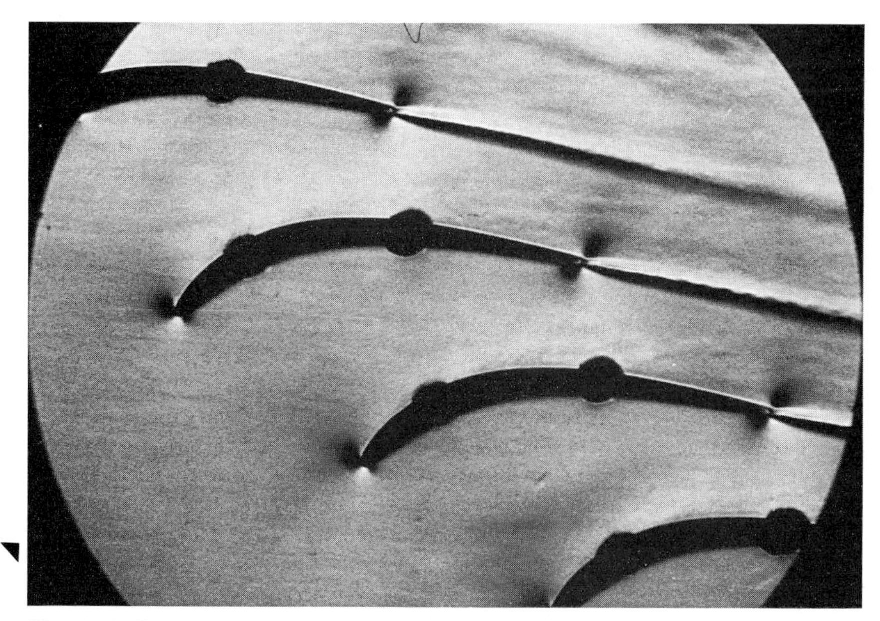

Fig. 176. Subsonic flow.

Fig. 177. Transonic flow.

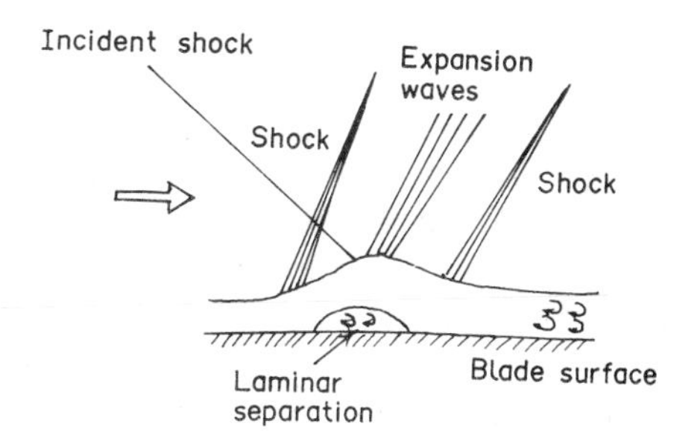

Fig. 178. Supersonic flow.

Air, outlet flow velocity (176) 233 m/s, (177) 441 m/s, (178) 527 m/s, blade chord 33.6 mm, Re = (176) 4.1×10^5, (177) 5.2×10^5, (178) 5×10^5, *schlieren method.*

Flow through a turbine cascade (from normal operation to the stalled state)

A turbine usually operates through a wide range, and it is important to know how the flow through the turbine rotor changes from normal operation to the stalled state.

Figures 179–181 show *interferograms* for different incidences.

Figure 179 shows normal operation at zero incidence, when the flow enters the cascade from the lower left and leaves to the right with no *separation*.

In Fig. 180, the flow entering from the left and leaving to the right separates at the leading edge on the pressure surface, but it recovers, to reattach again on the downstream half of the blade, due to the acceleration of the flow within the cascade.

In the case of larger negative incidence shown in Fig. 181, when the flow enters from the upper left, an enlarged region of separated flow is visible as an area with no fringe.

Figures 182 and 183 give schlieren pictures corresponding to Figs. 180 and 181. It can be seen that, as the incidence deviates from the normal range, the separated region spreads out and a weak shock is generated due to the blockage formed by the separated flow.

Flow through an impulse turbine cascade (effect of blade profile)

Figures 184 and 185 give *interferograms* for two different profiles of an impulse turbine cascade, in which the flow enters the cascade from the lower left and leaves to the lower right after turning through a large angle within the cascade.

As shown in Fig. 184, when the flow inlet angle deviates from zero incidence, *flow separation* occurs at the leading edge on the suction surface, leading to deterioration of the turbine performance.

Figure 185 is the case when the flow angle coincides with zero incidence, and shows no flow separation.

In general, the characteristics of an impulse turbine depend strongly on the flow angle, and the blade profile is designed to be tangential to the inlet flow on its suction surface.

Figures 186 and 187 are schlieren pictures corresponding to Figs. 184 and 185.

Figure 186 clearly shows the flow separation at the leading edge, whereas, in Fig. 187, no flow separation is visible, and the steady growth of the surface boundary layer up to the wake may be clearly seen.

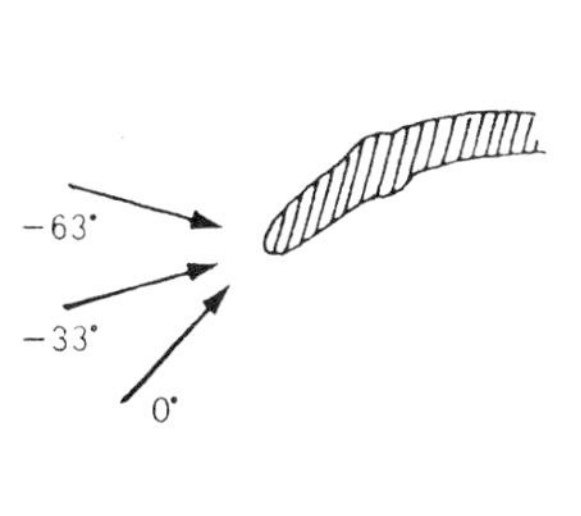

Fig. 179. 0° incidence.

Fig. 180. −33° incidence.

Fig. 181. −63° incidence.

Air, outlet velocity (179) 292 m/s, (180) 287 m/s, (181) 289 m/s, blade chord 33.6 mm, $Re = 4.8 \times 10^5$, Mach–Zehnder interferometry.

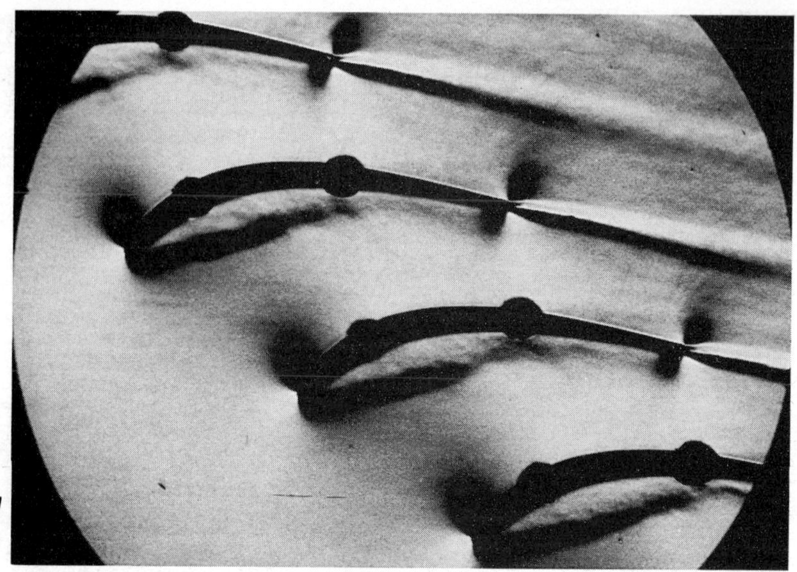

Fig. 182. −33° incidence.

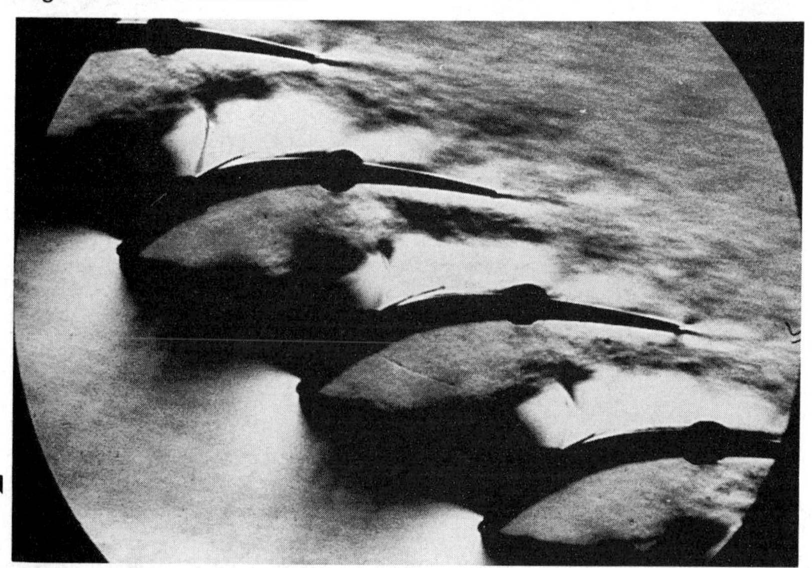

Fig. 183. −63° incidence.

Air, outlet flow velocity (182) 287 m/s, (183) 289 m/s, blade chord 33.6 mm, $Re = 4.8 \times 10^5$, schlieren method.

Secondary flow in an annular, radial inflow, cascade

Pronounced streamline curvature occurs in an annular cascade and this produces large angular deflections. In the main flow region, the centrigual force balances the pressure gradient. In the boundary layer regions on the side walls, however, this balance is upset because the flow velocity in the boundary layer is slower than that in the main flow region. This unbalance causes a secondary flow which turns from the pressure side to the suction side of the vane along the side walls of the cascade.

Figure 188 shows secondary flow on the side wall of an annular cascade which is installed in a return channel of a multi-stage *centrifugal compressor*. The flow is visualized by the *Oil-Film method*. The main flow enters into the cascade from the lower left. The secondary flow, mentioned above, is visualized clearly on the side wall. Such a

Fig. 184. Interferogram of flow through a turbine cascade with separation.

Fig. 185. Interferogram of flow through a turbine cascade without separation.

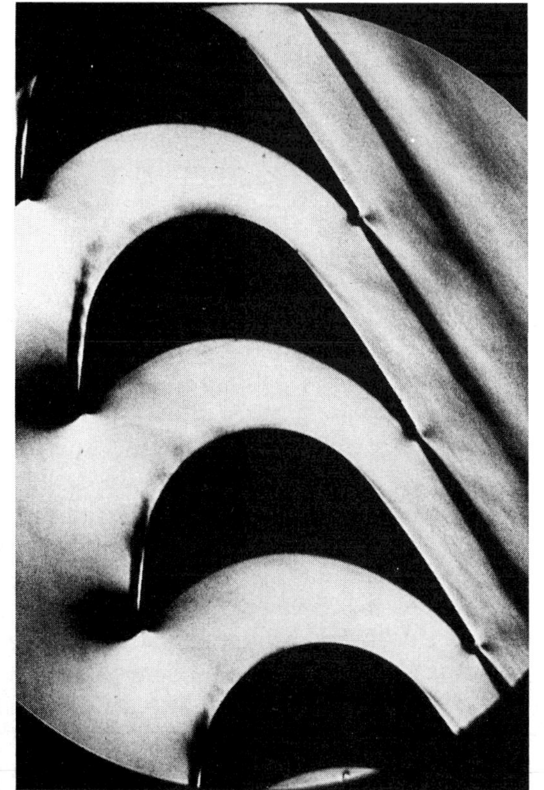

Fig. 186. Schlieren photograph of flow through a turbine cascade with separation.

Fig. 187. Schlieren photograph of flow through turbine cascade without separation.

Air, outlet flow velocity 232 m/s, blade chord 30 mm, inlet flow angle 72°, $Re = 4.4 \times 10^5$, (184, 185) Mach–Zehnder interferometry, (186, 187) schlieren method.

Fig. 188. Secondary flow in an annular cascade.

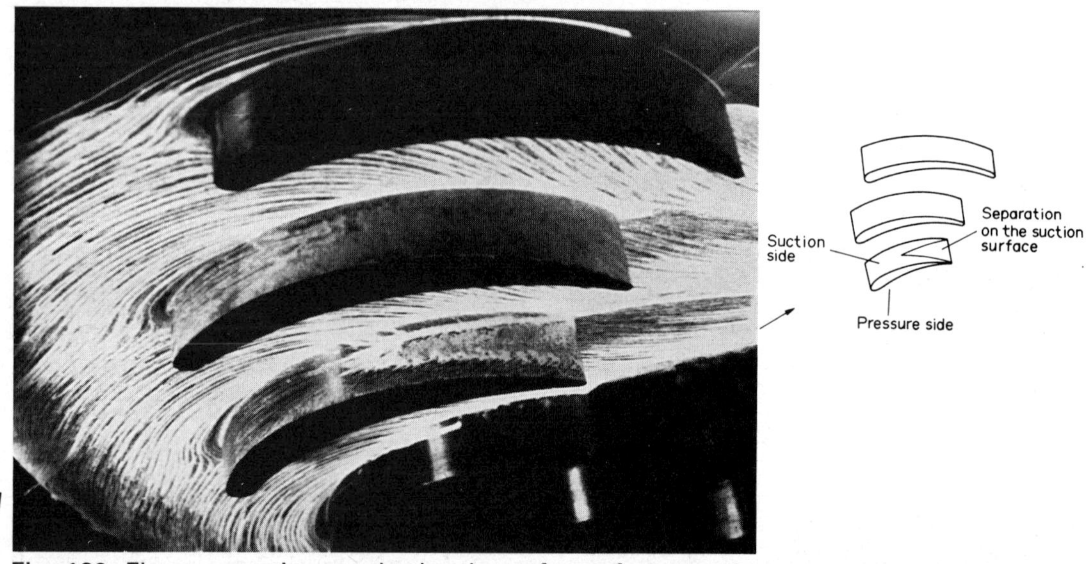

Fig. 189. Flow separation on the brade surface of an annular cascade.

Air, inlet flow velocity 40 m/s, vane chord 152 mm, $Re = 4 \times 10^5$, *oil-film method.*

secondary flow causes non-uniform flow conditions behind the cascade, and increases the loss of head. Further, *flow separation* is observed on the side wall in front of the leading edge of the vane.

Flow separation on a vane surface of a circular cascade

The condition of flow separation on the surface of a vane in an annular cascade, installed in a narrow channel, differs from that on a two-dimensional vane such as an isolated aerofoil.

This is particularly so on the suction surface of the vane. The secondary flow on the side walls (Fig. 188) brings about a flow toward the midspan of the vane along the suction surface, and flow passes back along the middle plane of the passage from the suction side of one vane to the pressure side of the next. That is, the flow separations on the vane surface near the side walls are suppressed because momentum is brought into the boundary layer. Whereas, at the midspan region, the induced receding flow promotes flow separation. Accordingly, a triangular separated region is formed on the

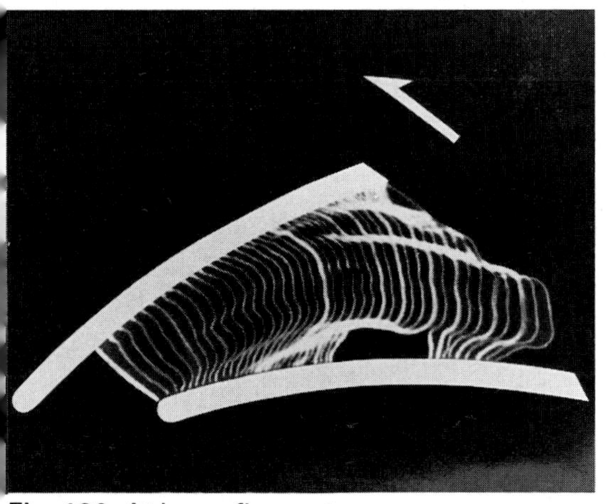

Fig. 190. At large flow rate.

Fig. 191. At design point.

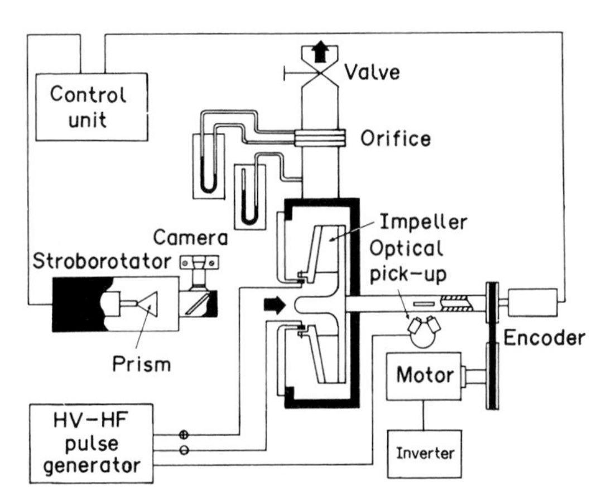

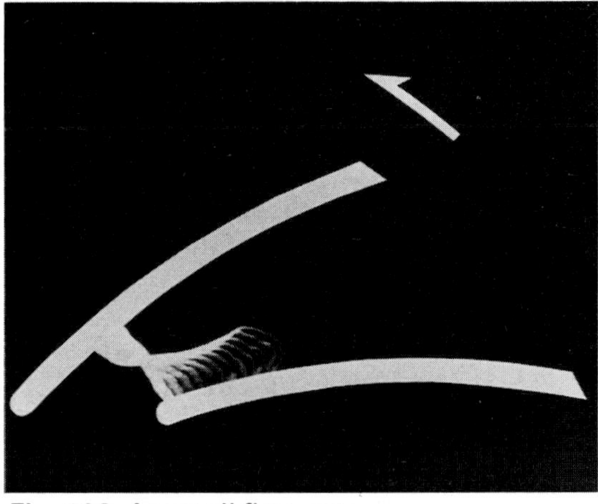

Fig. 192. At small flow rate.

Air, rotating speed 2500 rpm, impeller diameter 375 mm, $Re = 6 \times 10^5$, spark tracing method.

suction surface as shown in Fig. 189. In this area, some pressure recovery is obtained in the streamwise direction.

On the other hand, flow separation on the pressure surface is limited to a region near the trailing edge with no variation in the spanwise direction, though a flow from the midspan to the side walls is observed.

Flow through an impeller

The flow through a rotating impeller can be observed as if it were stationary by using a prism which rotates at one half of the impeller speed.

The sketch shows the scheme of the above method. The *spark-tracing method* is used here to visualize the flow. The electric rods are embedded on the midspan of the opposing surfaces of two neighbouring impeller blades, which are made of acrylic resin. By supplying an electric pulse of high frequency at high tension to the rods, a sequence of the *time-lines* is generated which move according to the flow.

Figure 190 shows the case at an excessively large flow rate. The flow velocity becomes higher on the suction side but separation occurs on the pressure side with reversed flow.

Figure 191 shows the case at the design flow rate. The *separation* region becomes very small and the flow leaves uniformly from the impeller along the exit angular direction.

Figure 192 is a picture taken at a time of exceedingly small flow rate, and reverse flow due to surging is clearly visible.

Unsteady Flow

Pulsating flow

In Fig. 193, smoke profiles of pulsating flow in a rectangular channel are shown in a series every $\pi/4$ radians. The smoke profiles show only approximate velocity profiles in the present case.

At the end of the channel a harmonic flow is generated by the motion of a piston. The other end is open to the atmosphere. A thin nichrome wire is strained across the channel at the middle of the test section.

The velocity of the flow reaches its maximum near the solid boundary. This is the well known *Richardson annular effect*.

Flow around a rotatory-oscillating cylinder

The boundary layer on a circular cylinder performing a rotatory oscillation in still water forms a series of ring-shaped vortices (Fig. 194(a)). The vortices grow with time and are finally broken down. The manner of development of the vortices and their subsequent history is similar to that for a circular cylinder started rotating in still water (Fig. 194(b)). The formation of these ring-shaped vortices is ascribed to centrifugal instability.

Laminar boundary layer flow on an oscillating flat plate

Figure 195 shows a series of smoke profiles on an oscillating flat plate immersed in a uniform flow. A thin nichrome wire is strained normal to the plate at a point 25.4 cm downstream from the leading edge. The tip of the probe is lightly in contact with the plate, but does not move with it. (a) shows a smoke profile at the instant when the oscillating plate has maximum speed to the right (b) through (d) show smoke profiles whose phases are delayed by $\pi/4$ radians when compared with the previous one.

The motion of the plate can be inferred from the profile in the immediate neighbourhood of the plate. If the plate were stationary, the *Blasius velocity profile* would be observed.

The present flow, whose velocity profile changes periodically, is of interest in connection with the study of the transition of a boundary layer flow from laminar to turbulent on a flat plate.

Sloshing of two-layered liquids within a rectangular container

Streamlines of the sloshing are shown in Figs. 196 and 197. The liquids, i.e. kerosene in the upper layer and water in the lower layer, are oscillating, forced by sinusoidal displacement of the container in the horizontal direction. The density ratio is 0.79 and the depth of each layer is 100 mm.

The pictures show respectively the modes at frequencies close to the first fundamentals of the resonance arising on the free surface and on the interface separating the liquids, which are 0.38 Hz and 1.20 Hz respectively. The streamlines calculated from the equation of potential flow, assuming an inviscid flow and a small amplitude oscillation, agree well with the experimental observations.

The tracers are particles of foamed polystyrene for the kerosene, and for the water a mixture of toluene and carbon tetrachloride, with a density equal to the water density. The pictures are taken by a camera moving with the container, after damping out of transient waves arisen at the beginning of the excitation. The left half of the field is a complete mirror image of the right half with respect to the centre line.

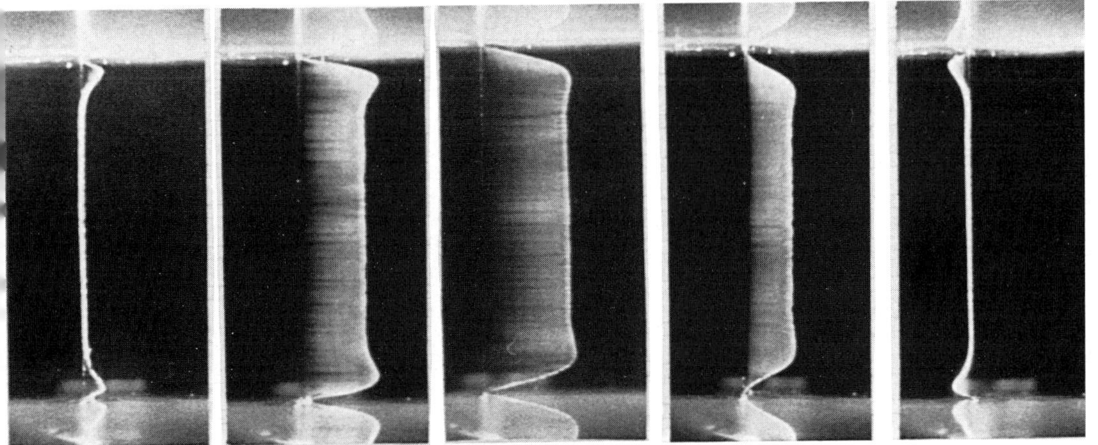

Fig. 193. Air; maximum instantaneous speed of piston 16 cm/s; rectangular cross section 50 mm (across the channel) by 150 mm (perpendicular to the page); channel length 4300 mm; implitude of the piston movement 38 mm: frequency 1 Hz; smoke wire technique.

Standing flow pattern induced by an oscillating cylinder

A cylinder, oscillating in a liquid in a direction perpendicular to its axis, induces a superharmonic oscillation of flow superposed with a steady flow. At low Reynolds number, particularly Fig. 199 where $Re = 35$, eight vortices appear in a symmetric arrangement: as shown in the inset four vortices each on the exterior and in the interior of a circular ring concentric with the cylinder. However, in Figs. 198 and 199, the interior vortices are not as clear as the exterior vortices.

A stroboscopic flash light is used for the illumination, synchronized with a phase during the oscillation such that the centre of the cylinder is passing the origin in the forward to the left or to the right direction. So that the picture shows the trajectories of particles in the synchronous phase.

At increased Reynolds number, $Re = 2.1 \times 10^3$ in Fig. 198, a pronounced outgoing flow appears extending in the direction of the oscillation from the cylinder surface.

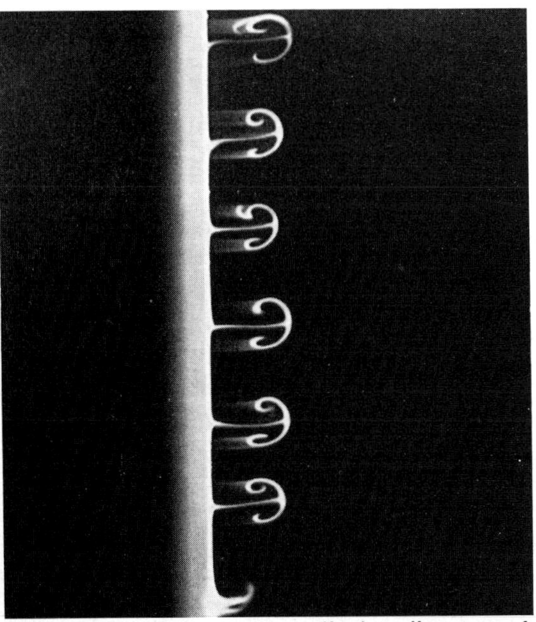

Fig. 194(a). Still water, cylinder diameter 1 cm, frequency 0.1 Hz, angular amplitude 270°, electrolytic precipitation method.

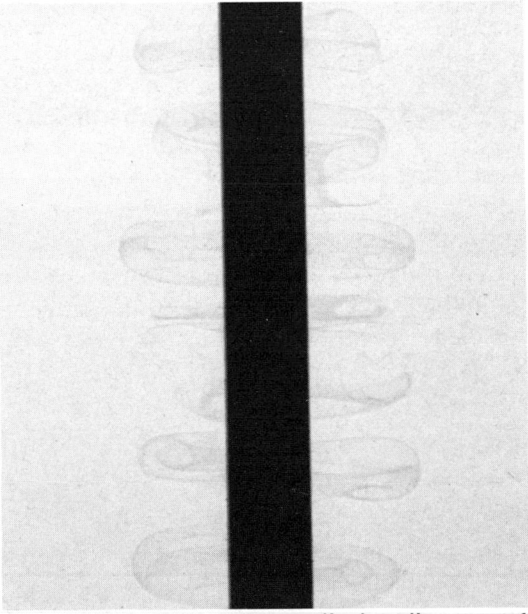

Fig. 194(b). Still water, cylinder diameter 1 cm, number of revolutions rps, electrolytic pH indication liquid method.

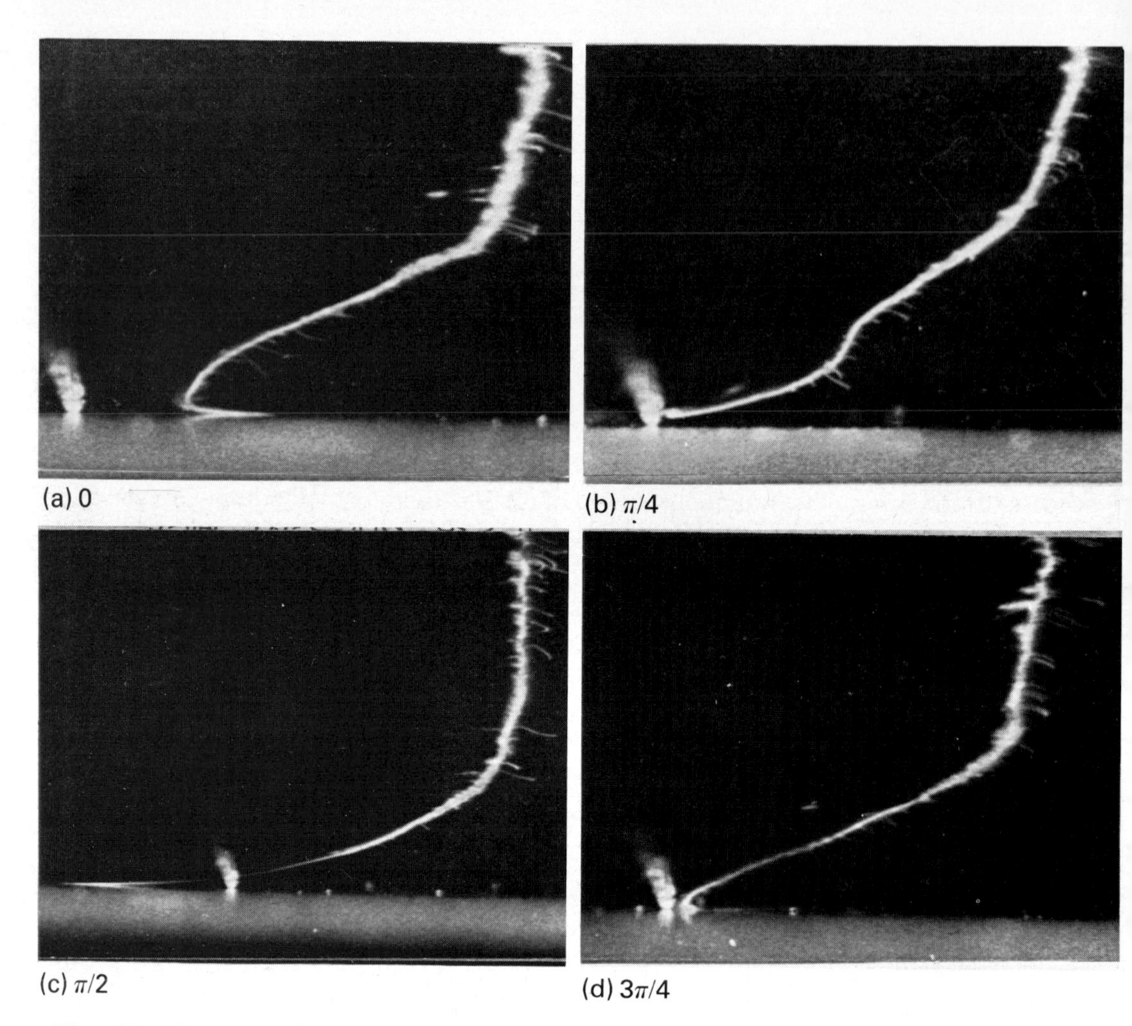

(a) 0 (b) $\pi/4$ (c) $\pi/2$ (d) $3\pi/4$

Fig. 195. Air, air speed 43 cm/s, oscillation amplitude of plate 25.4 mm, frequency = 2Hz; $Re = 7.3 \times 10^3$, smoke wire method.

However, the *standing pattern of vortices* and the concentric circular ring is hardly sustained as seen in Fig. 198. This is partly because the reduced viscous damping is unable to sustain the stable standing pattern, and partly because the superharmonic modes of the fluid response are intensified in the oscillating field.

Standing flow pattern around an oscillating flat plate

The flow patterns shown here are obtained using an illumination method similar to that used in the previous section. The flat plate with squared edges is oscillating sinusoidally in the longitudinal direction in a glycerine solution.

At a Reynolds number of 153, the pattern of vortices shown in Fig. 200 is formed, with a *pair of vortices* at each side of the plate and an additional pair of minor vortices against each plate end.

As the Reynolds number is increased to 304, shown in Fig. 201, another pair is formed at each end of the plate. Since the displacement is not exactly parallel with the plate, the arrangement of these vortices is slightly skewed with respect to the centre line of the plate. The minor vortices which appeared against the ends at the Reynolds number of 153 have been displaced to the side surfaces of the plate, so that the total number of vortices is six on each side of the plate.

With further increase in Reynolds number, the stable flow pattern becomes disturbed and the path-lines forming the vortices are hardly discernible.

Fig. 196. Streamlines showing sloshing at the frequency of the first fundamental mode of the free surface.

Fig. 197. Sloshing at the frequency of the first fundamental mode of the interface (kerosene/water, exciting frequency 2.5 Hz, full displacement amplitude 20 mm, container: 400 mm length, 100 mm width and 600 mm depth, tracer method).

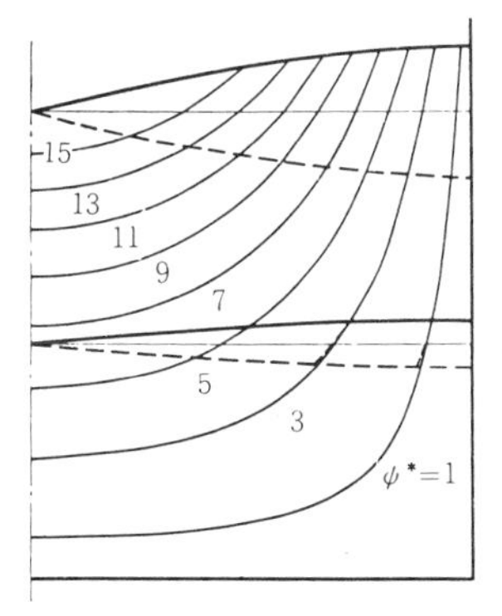

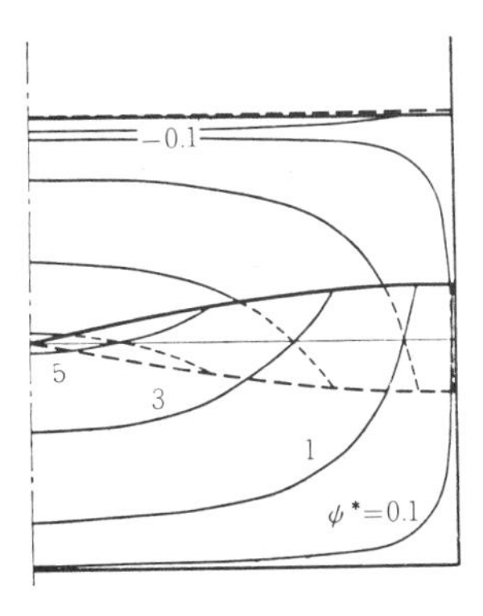

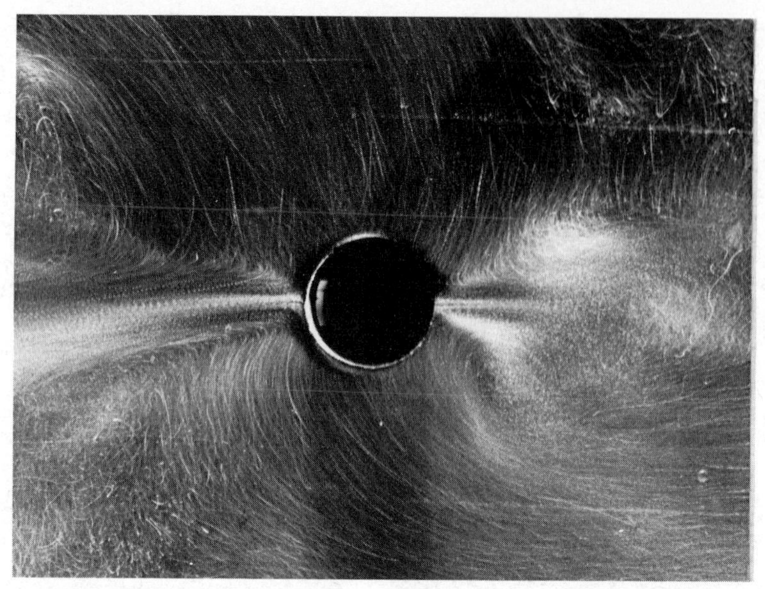

Fig. 198. Flow excited by an oscillating cylinder in water.

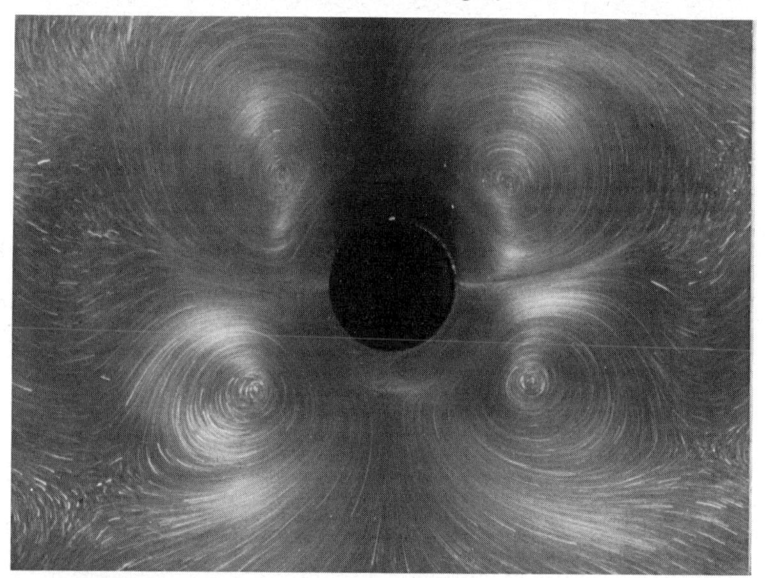

Fig. 199. Flow excited by an oscillating cylinder in glycerine [water (198), glycerine (199), excitation frequency 3 Hz, amplitude of oscillation 4.1 mm, diameter of the cylinder 30 mm, $Re = 2.1 \times 10^3$ (198), 35 (199), aluminium tracer method].

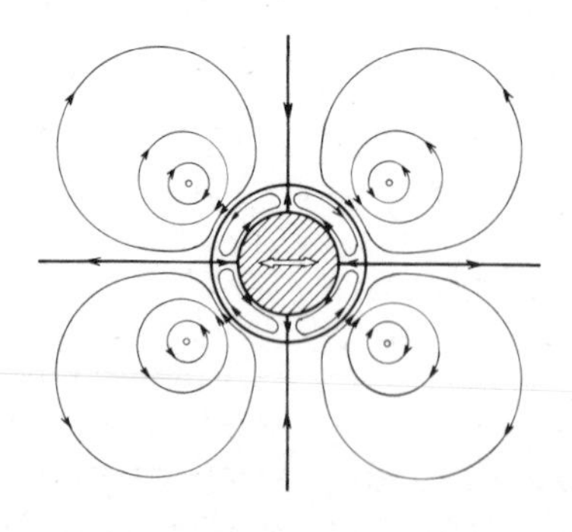

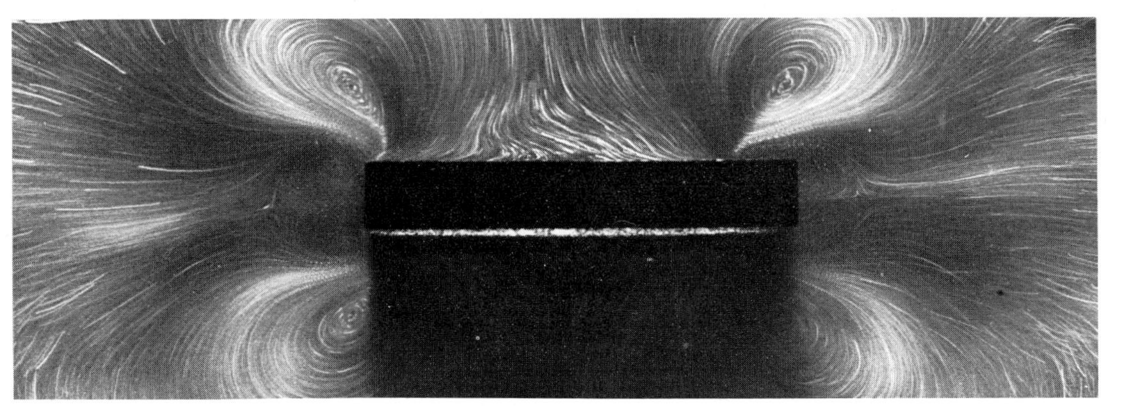

Fig. 200. Vortex pattern generated by longitudinal oscillation of a plate at 3 Hz.

Fig. 201. Vortex pattern at 5 Hz.

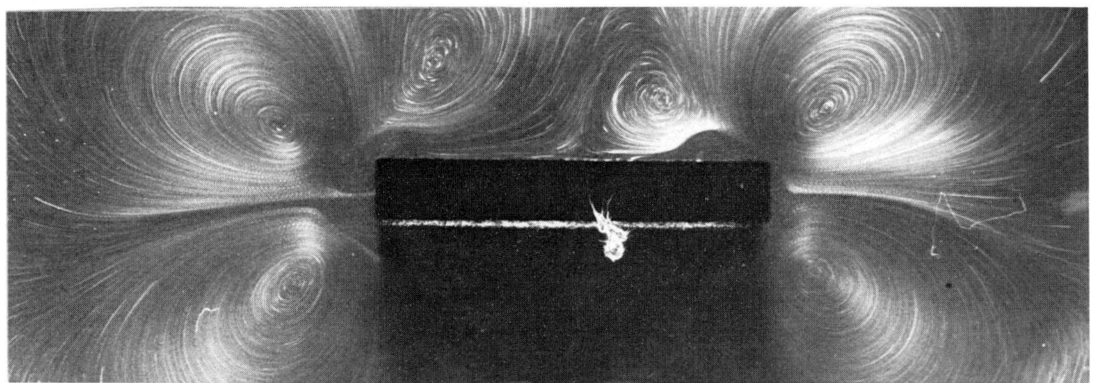

Glycerine, frequency (200) 3 Hz, (201) 5 Hz, amplitude of oscillation 5.4 mm, chord length 100 mm, plate thickness 15 mm, *Re* = (200) 153: (201) 304, aluminium tracer method.

Water waves

When a wave is generated on the surface of a liquid such as water, the particles of liquid move along certain curves. In the two-dimensional case, when the liquid is deep enough, the particles move along exact circles. The diameter of the circle is maximum at the surface and smaller as the position of particle is deeper. In this case, shown in Fig. 202, the wave is called a *deep water wave*.

In the shallow condition, the particles of liquid move along elliptical paths and the ratio of semimajor axis to semiminor axis becomes larger as the position of the particle is deeper. The movement of a particle at the bottom is a linear reciprocation. In this case, the wave is called a *shallow water wave*.

Figure 203 shows path lines traced out by resin particles which are suspended in the water of a tank. The resin has the same density as the water. The wave is generated from an end wall of the tank, and the photograph is taken with an exposure time nearly equal to the wave period. Because of the long exposure time, it is not possible to photograph the wave profile directly. So for reference, an instantaneous wave profile is drawn above the photograph as an addition.

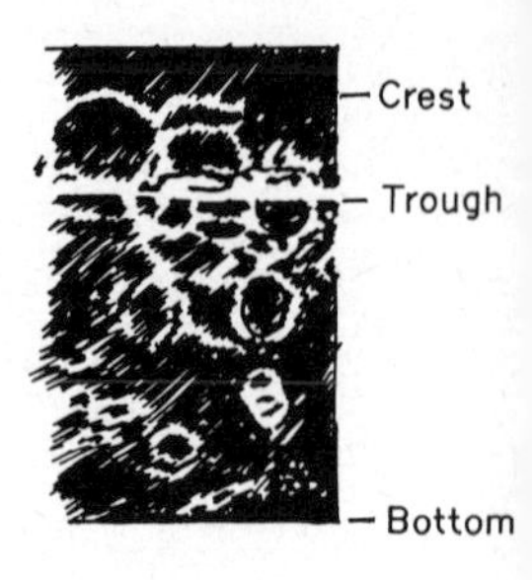

Wave profile at water surface

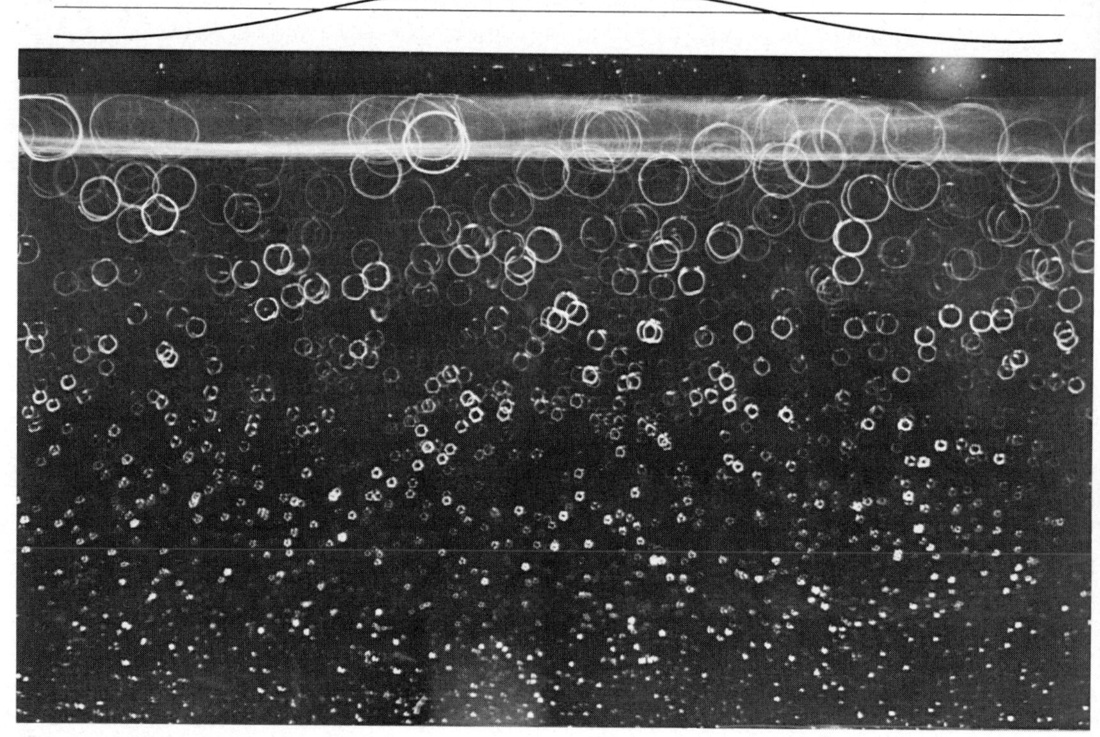

Fig. 202. Wave in deep water.

Wave profile at water surface

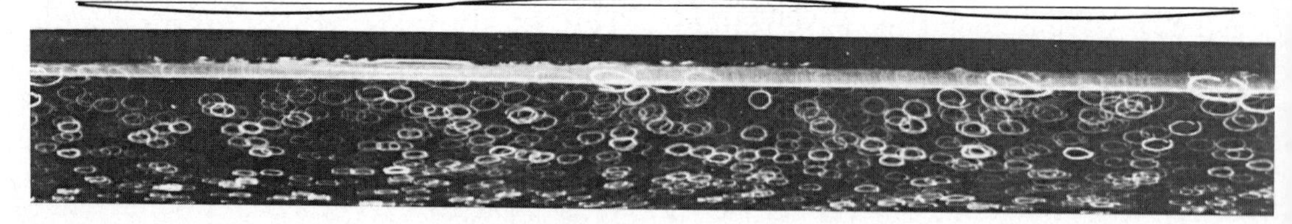

Fig. 203. Wave in shallow water.
Aqueous sugar solution, wave length, deep water = 200 mm, wave length, shallow water – 150 mm, solid particle suspension method.

Cavitation

Cavitation in nozzle and orifice

When the pressure on the surface of a body moving in a liquid, or on the surface of a channel or a pipe in which a liquid is flowing, or within a liquid itself, decreases below the vapour pressure, the liquid tends to form cavities which expand and relieve the low pressure. The cavities are carried downstream to a region of higher pressure and there collapse into small bubbles. The phenomenon on the inception, development and collapse of such cavities is called *cavitation*.

Figure 204 shows the development of the cavities formed in the throat of a *venturi tube*, whose shape and dimensions are shown in the auxiliary figure, with decreasing *cavitation number* $k_d = 2(p_\infty - p_v)/\rho U_\infty^2$, where p_∞ and U_∞ are the absolute static pressure and velocity at a point corresponding to infinity, and p_v and ρ are respectively the vapour pressure and density of water at the prevailing water temperature. In this case, k_d is calculated using the mean velocity in the throat as U_∞ and the static pressure measured on the wall of the throat as p_∞. Under the continuous lighting of an incandescent lamp, as shown in the left-hand column of Fig. 204. Cavitation can be seen as the region that has a vague boundary and is apparently filled with a milk-white substance. With a strobe light, as shown in the right-hand column of Fig. 204, individual cavities formed can be clearly seen in the cavitating region.

Figure 205 shows cavitation occurred in the region of high shear at the boundary between a submerged jet from an *orifice plate*, and surrounding fluid. It is deduced from the cavitation in the photograph, taken at a high frame rate and at $k_d = 0.13$, that vortex rings are being shed from the edge of the orifice plate perpendicularly to the jet stream and are breaking down into a cloud of tiny bubbles as they go downstream. Here, k_d is calculated using the mean velocity through the orifice and the static pressure measured at the downstream pressure tap of vena contracta taps.

Cavitation on the surface of an isolated hydrofoil

Cavitation is classified into two types, which are termed *travelling cavitation* and *fixed cavitation*. It is generally accepted that the inception of cavitation is associated with the growth of submicroscopic nuclei containing vapour, undissolved gas, or both. Travelling cavitation is a type composed of individual cavities or bubbles which grow from the nuclei and move with the water at the speed nearly equal to the free stream speed as they expand. They begin to shrink, when they reach a region where the pressure exceeds the vapour pressure, and then collapse, sometimes rebounding several times before collapsing completely.

Figure 206, obtained using continuous lighting with an incandescent lamp (at the left) is an example of *travelling cavitation*, which appears as if a gaseous pocket of milk-white vapour is attached to the surface. With a strobe light many bubbles can be seen to be moving downstream along the surface of the hydrofoil. *Fixed cavitation* is characterised by the cavity remaining attached to the body on which cavitation has been induced. The attached or fixed cavity, which is stable on a temporal average, is formed by the solid surface and the free surface of the water flow as shown in Fig. 207. Its instantaneous appearance is very like the temporally average

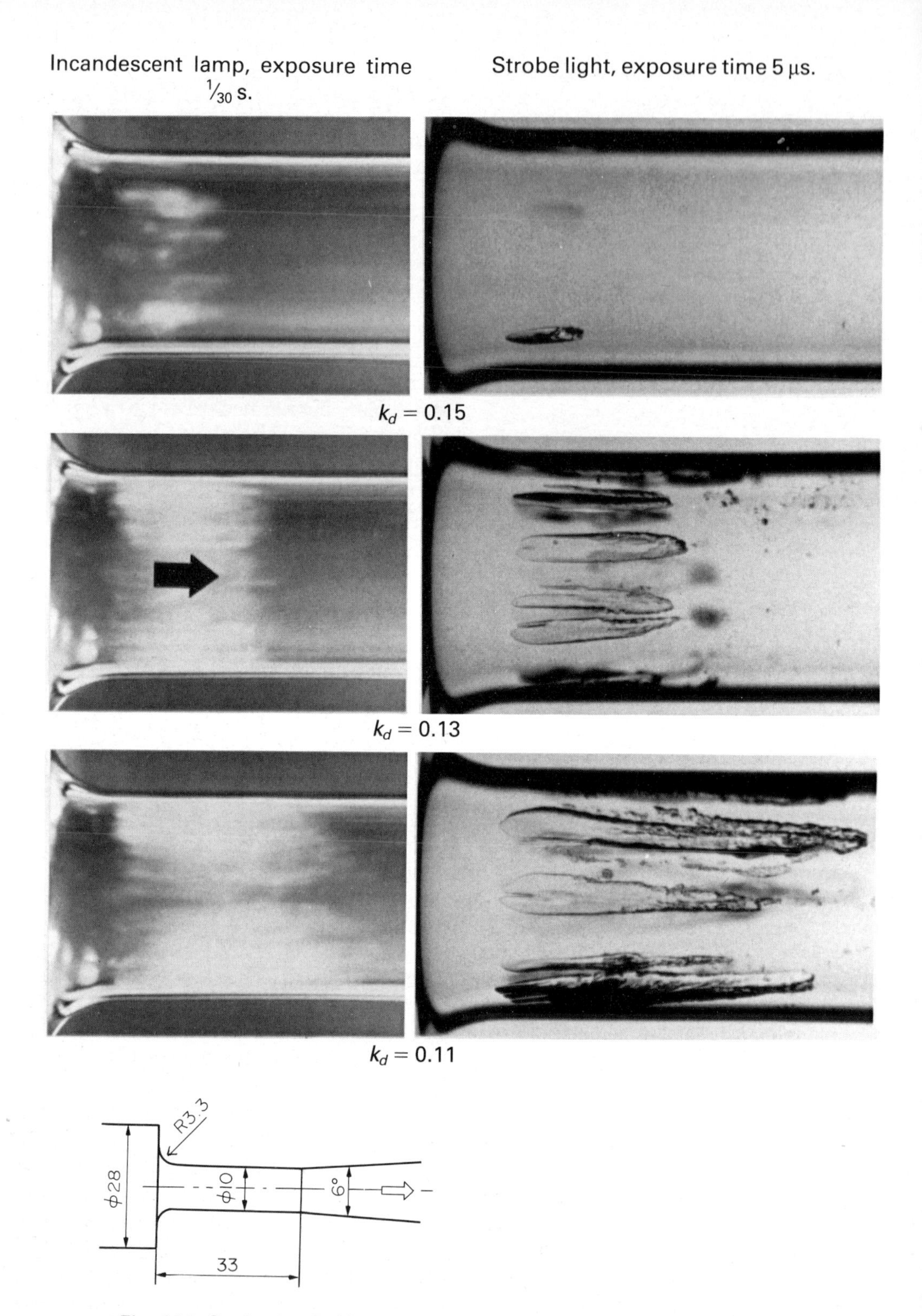

Fig. 204. Cavitation in Venturi nozzle (water, 15 m/s, throat diameter 10 mm, $Re = 1.4 \times 10^5$).

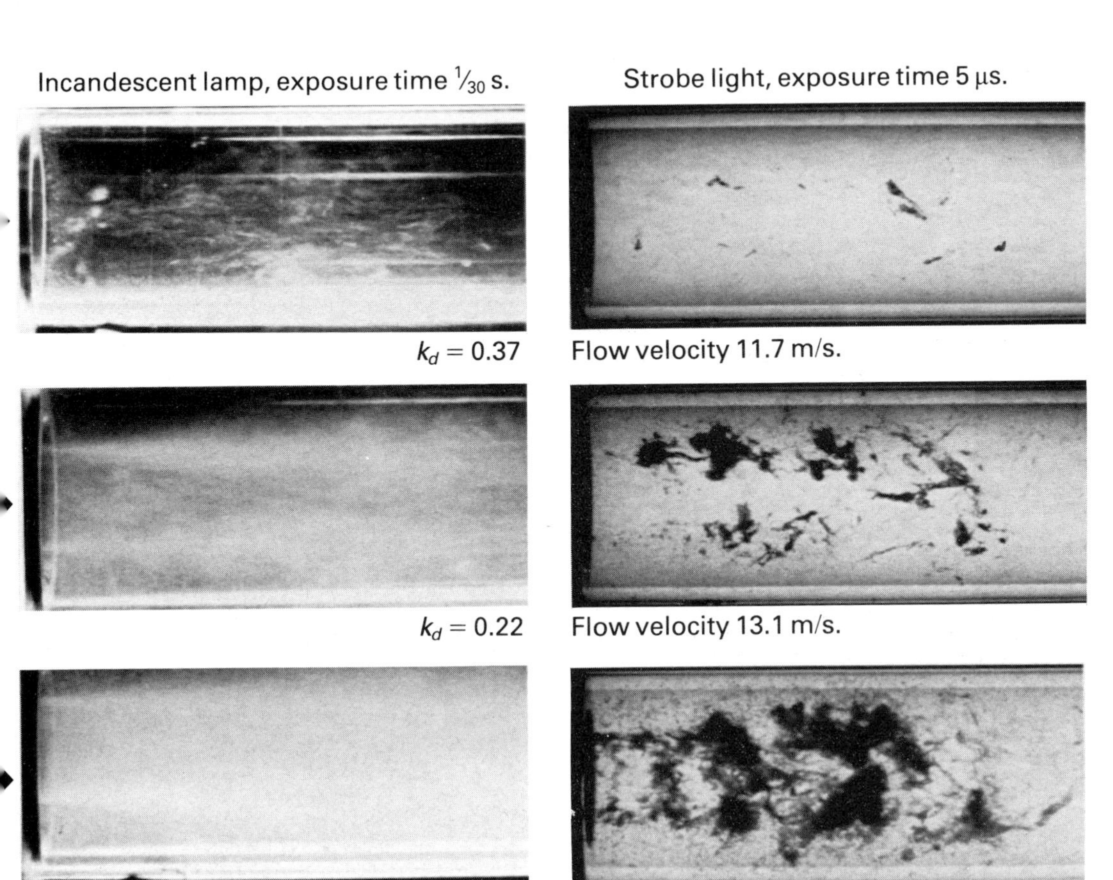

Fig. 205. Cavitation in an orifice gauge (water, 11.7 to 14.6 m/s, orifice diameter 10 mm, $Re = 1.2$ to 1.5×10^5).

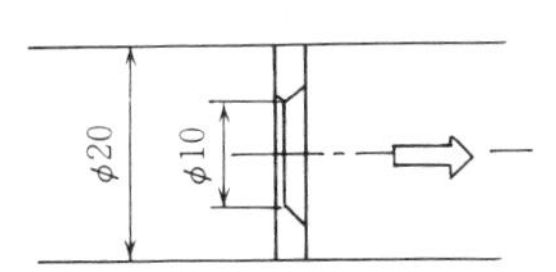

one except near the trailing end at which the cavity collapses.

The type of cavitation depends on many factors, such as the shape of the body, surface roughness, size and number of nuclei, dissolved and total gas content, etc. It frequently happens that both types are observed simultaneously, when travelling cavitation develops, after inception individual cavities combine together downstream of their first appearances and form a single cavity. Fixed cavitation is observed as a single cavity even at the incipient stage.

As k_d is lowered the length of the cavity increases and then fluctuates rather rapidly. The trailing end of the cavity exhibits quite an unstable motion, leading to oscillating forces that cause rather severe vibrations of the hydrofoil and induce violent noises. The surface of the hydrofoil is subject to severe shock pressure and damage when the trailing end of the cavity collapses on to the surface. The oscillations of the cavity die out when its length exceeds about 120% of the hydrofoil chord length. Under such a condition the flow around the hydrofoil and cavity becomes stable, and is called *supercavitating flow*.

Incandescent lamp, exposure time $^1\!/_{25}$ s. Strobe light, exposure time 5 μs.

$k_d = 1.30$.

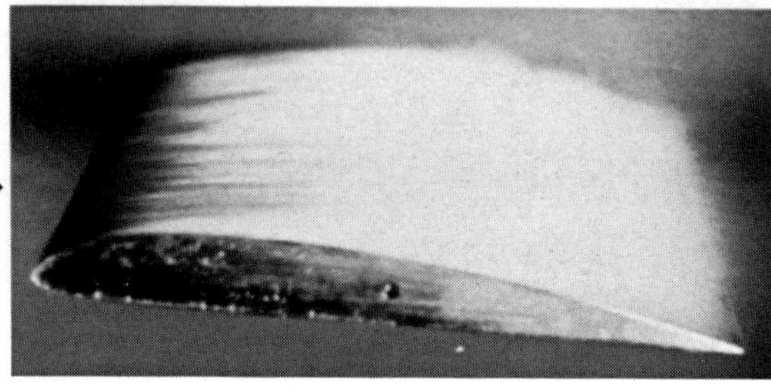
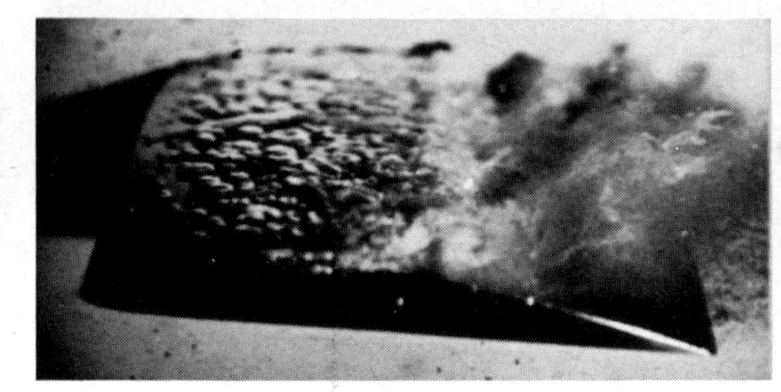

$k_d = 1.09$.

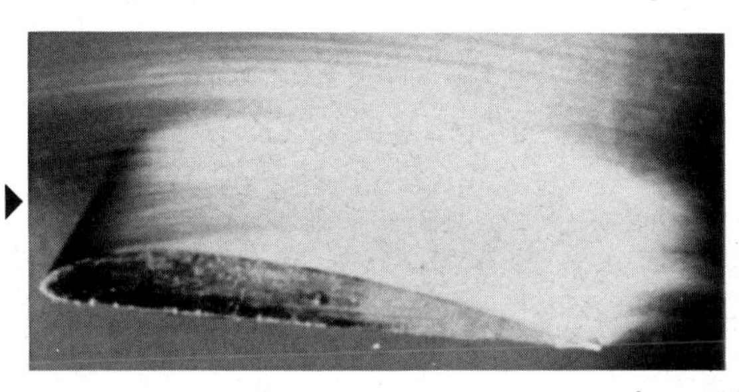

$k_d = 0.88$.

Fig. 206. Cavitation on a Clark Y-11.7% profile form (water, 11.4 m/s, chord length 100 mm, $Re = 1.1 \times 10^6$).

Cavitation behind a circular cylinder

Figure 208 shows cavitation behind a circular cylinder placed perpendicularly to a flow. When cavitation does not develop sufficiently to alter the flow pattern (at $k_d = 1.33$), cavities are observed to occur in the cores of the periodic vortices being shed alternately from the upper and lower surfaces. Being distinguished from travelling and fixed cavitation, this type is called *vortex cavitation* and belongs to the same category as the cavitation which occurs behind an orifice plate as shown in Fig. 205. As k_d is lowered large cavities are shed, and forces acting upon the circular cylinder oscillate violently in the directions normal and parallel to the flow, causing severe vibrations. With further decrease of k_d a steady fixed-cavity is formed (at $k_d = 0.90$). Here, vortex cavitation can also be seen to occur in the necklace vortex generated in the neighbourhood of the side wall.

Collapsing mechanism of a cavitation bubble

When a *bubble collapses* close to a solid boundary, owing to a pressure rise in the surrounding water, the bubble surface becomes unstable and *a microjet* impinging on the boundary is formed as shown in Fig. 209. If two bubbles collapse very close to each other the directions of their microjets will be changed due to interaction. *A shock*

Incandescent lamp, exposure time $\frac{1}{25}$ s. Strobe light, exposure time 5 μs.

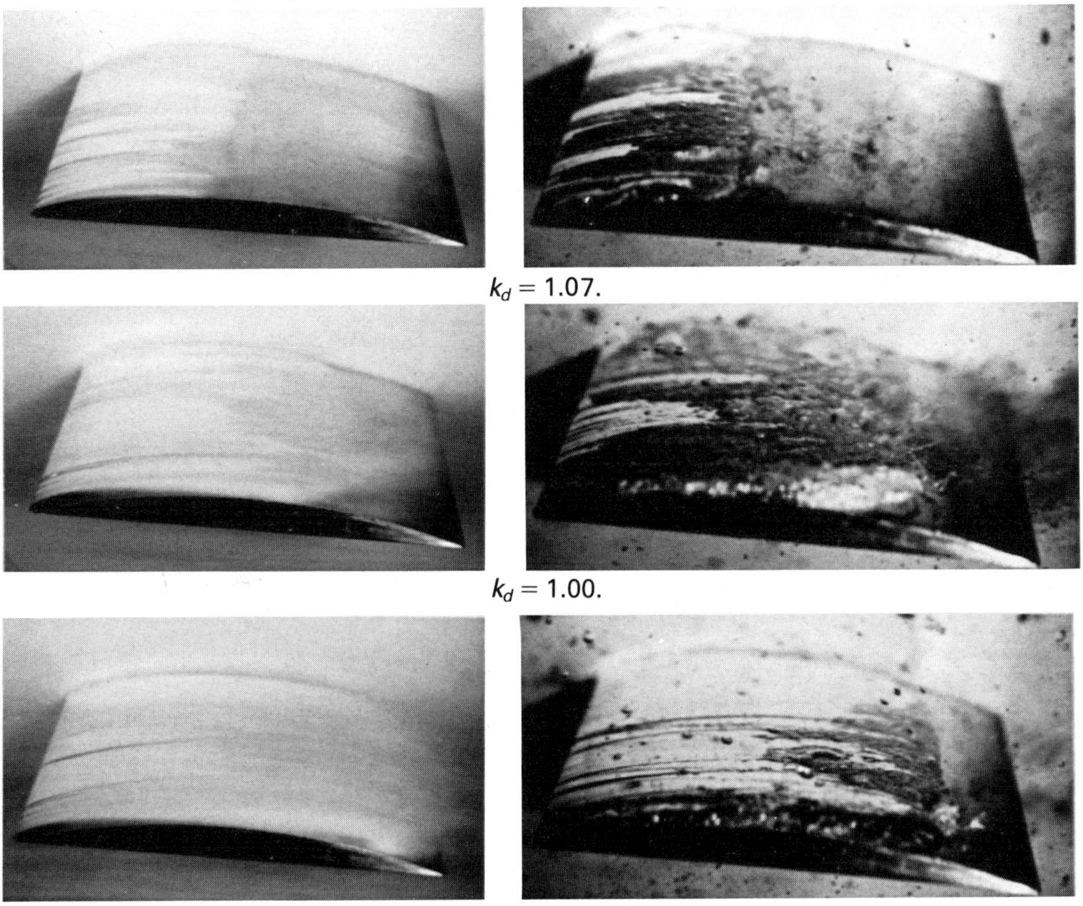

$k_d = 1.07$.

$k_d = 1.00$.

$k_d = 0.87$.

Fig. 207. Cavitation on an 8% thick profile form (water, 11.4 m/s, chord length 100 mm, $Re = 1.1 \times 10^6$).

released from a collapsed bubble, which is transmitted in the solid boundary as a stress wave. *Cavitation damage* is explained as being caused by the shock wave, or the impingement of the microjet, or both.

Cavitation in an axial flow pump

Figures 210–212 show the development of cavitation on an impeller blade of *an axial flow pump*. Each set of pictures is taken simultaneously using electronic flash illumination and two cameras arranged to take side and front views of the suction surface of the impeller blade, which is rotating counter-clockwise. The flow rate is 94% of the best efficiency point (b.e.p.). In the case of a pump having a non-shrouded impeller, such as an axial flow impeller, cavitation also occurs in the clearance between the blade tip and the casing. Pump performance is not reduced immediately with the inception of cavity but eventually deteriorates when severe cavitation has developed and has formed large cavities on the blade surface.

Cavitation in a mixed-flow pump

Cavitation in a *mixed-flow pump* is shown in Figs. 213–215. Water is flowing from the right to the left. The impeller is rotating counter-clockwise, and is being observed from the suction side (from the right). The flow rate at which these pictures were taken

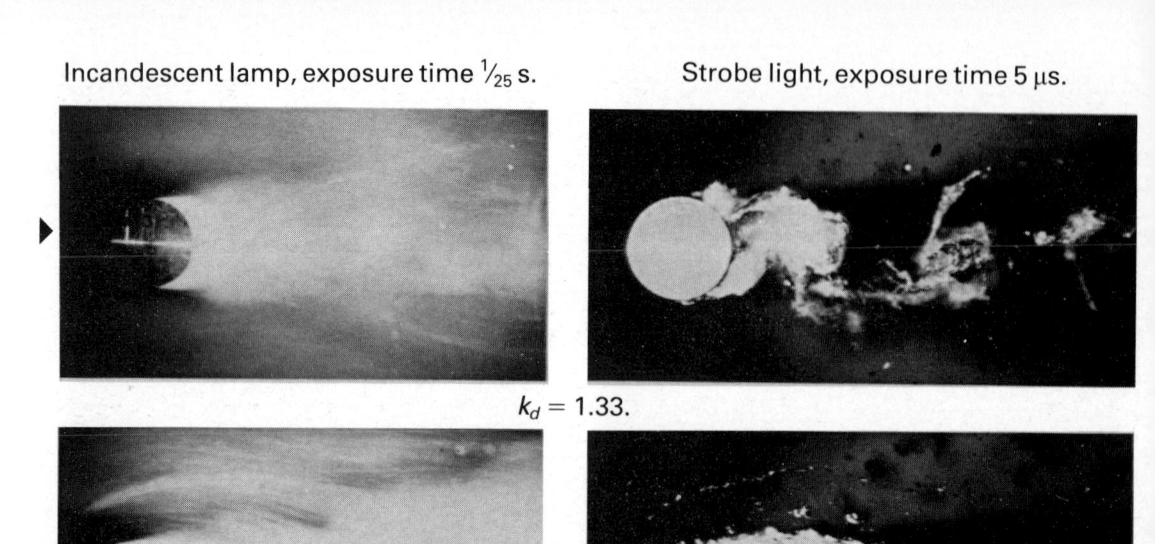

Fig. 208. Cavitation behind a circular cylinder (water, 11.3 m/s, cylinder diameter 18 mm, $Re = 2 \times 10^5$).

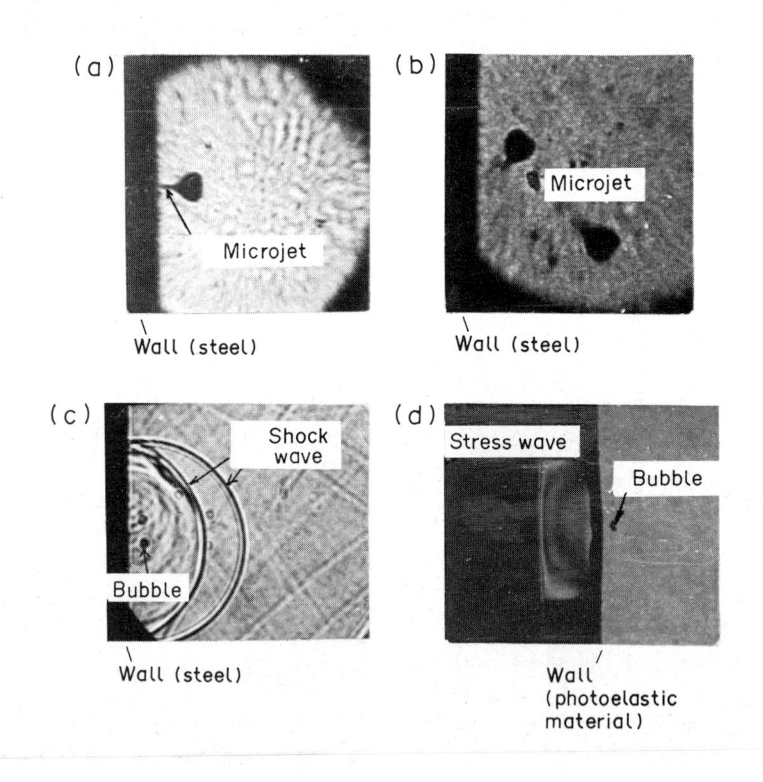

Fig. 209. Bubble(s) collapsing close to a solid boundary [filtrated water, $L/R_0 = 1.89$ (a), 1.16 (c), 1.65 (d); R_0: initial radius of the bubble, L: distance from the centre of the bubble to the solid boundary].

Fig. 210. Immediately after inception.

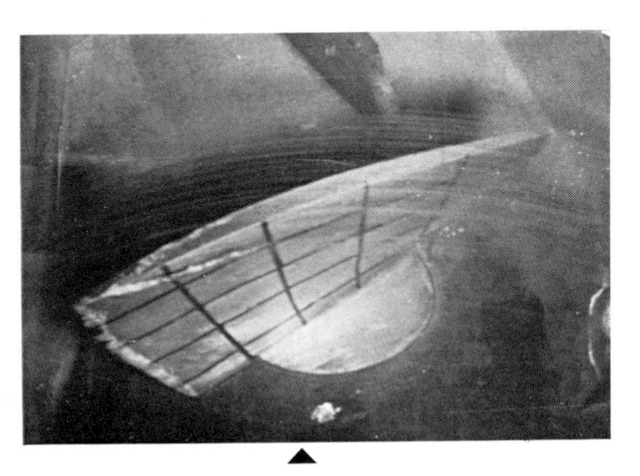

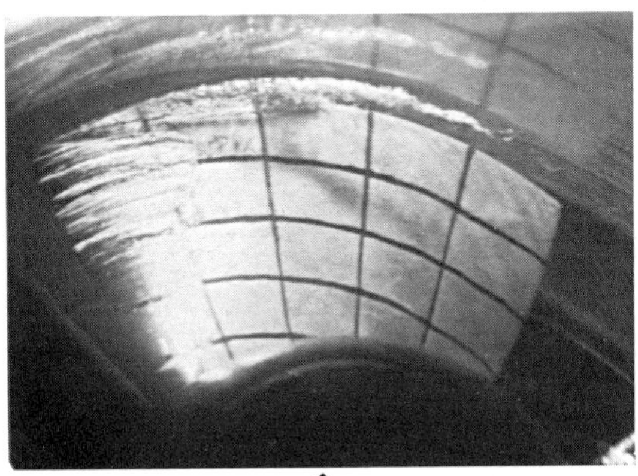

Fig. 211. At the beginning of efficiency reduction.

Figs. 212. At the beginning of head reduction.

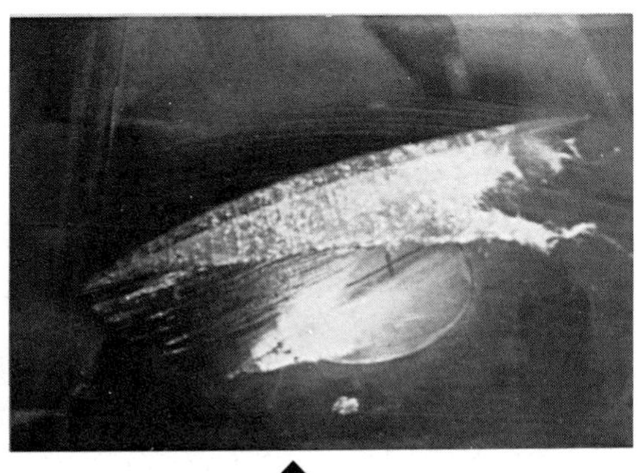

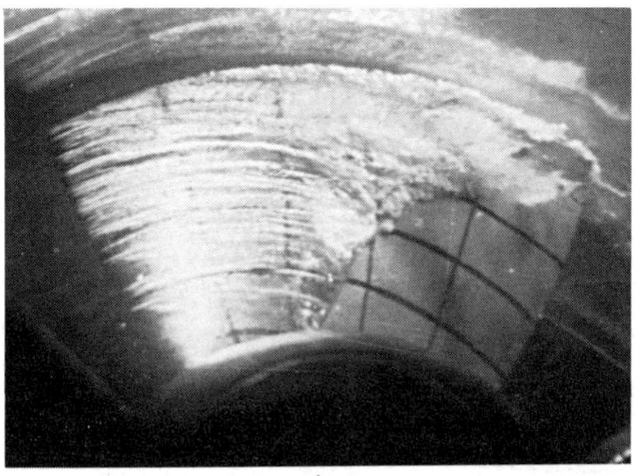

Figs. 210–212. Cavitation in an axial flow pump (water, rotational speed 1000 rpm, impeller diameter 356 mm).

is 140% of the b.e.p. At this excessive flow rate, cavitation occurs on the delivery side of the impeller blade because the flow entrance angle is larger than the blade angle. The severity of the cavitation, i.e. the cavity length, increases toward the blade tip because the circumferential velocity of the blade increases in proportion to its radius from the pump axis and, therefore, the relative velocity of the water increases, too (cf. Figs. 214 and 215).

Figures 216–218 show cavitation in the same pump at a flow rate of 70% of the b.e.p. Cavitation is observed to occur on the suction side by similar reasoning to that given above (cf. Fig. 216). The camera position is moved more toward the suction side than in the above case. Tip clearance cavitation can also be seen since this impeller is an open type,

Figs. 213–215. Cavitation in a mixed-flow pump at flow rate over normal (water, tip speed at impeller inlet 11.1 m/s, inlet diameter 245 mm).

Fig. 213. Immediately after inception.

Fig. 214. At the beginning of head reduction.

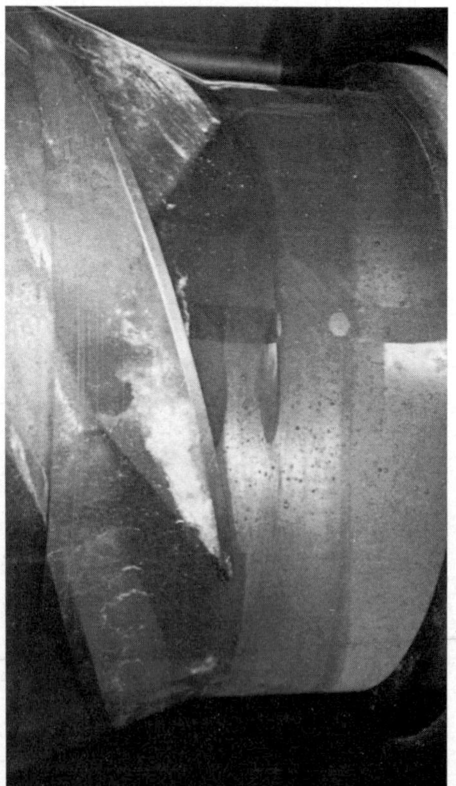

Fig. 215. At the point of sudden head reduction.

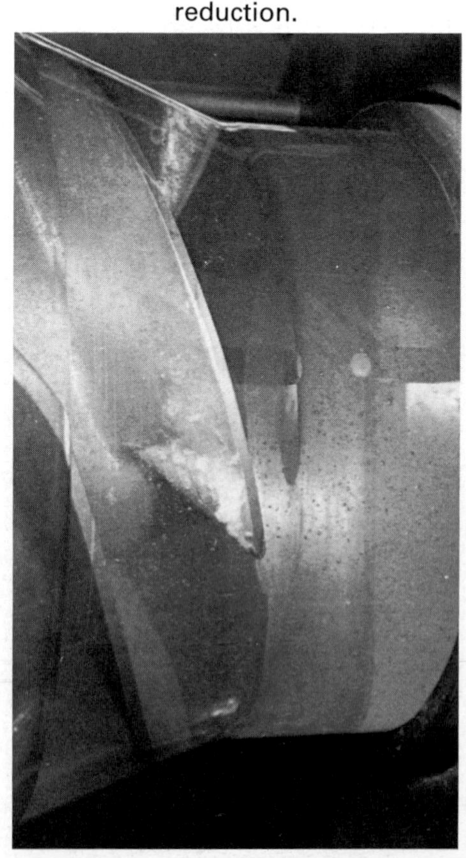

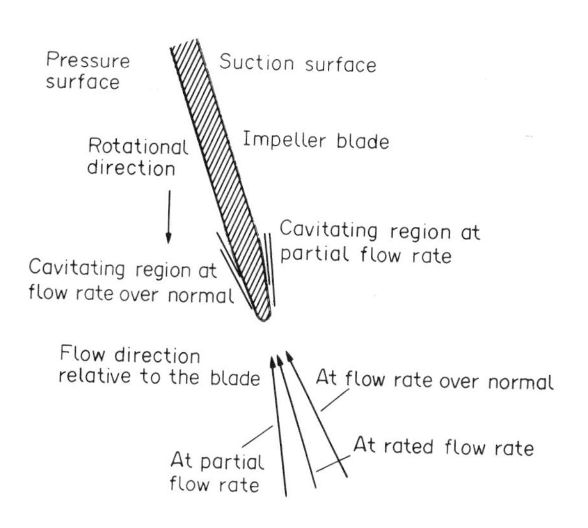

Figs. 216–218. Cavitation in a mixed-flow pump at partial flow rate (water, tip speed at impeller inlet 11.1 m/s, inlet diameter of impeller 235 mm).

Fig. 216. Immediately after inception.

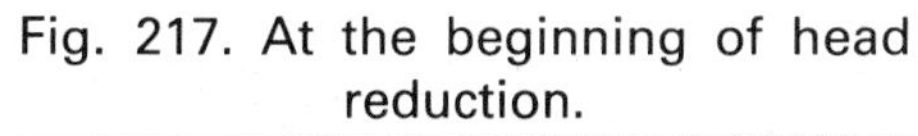

Fig. 217. At the beginning of head reduction.

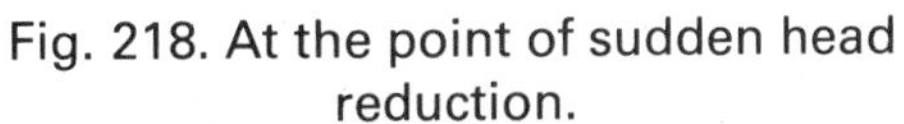

Fig. 218. At the point of sudden head reduction.

Fig. 219. Cavitation damage on the blade surface of a runner of a Francis turbine (water, tip speed 20 m/s, outlet diameter 1000 mm).

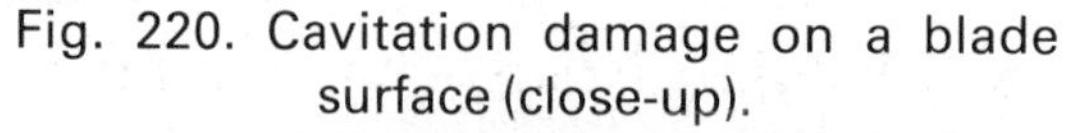

Fig. 220. Cavitation damage on a blade surface (close-up).

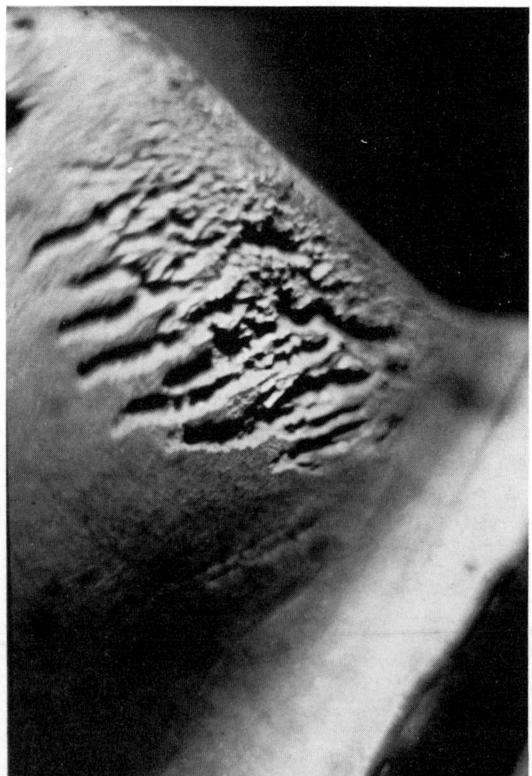

similar to the axial-flow impeller in the previous section. This tip clearance cavitation does not occur over the entire tip surface unlike the previous case of the axial-flow impeller because the chord length of this impeller blade is very much greater.

Cavitation damage on a Francis turbine runner

Figures 219 and 220 show a part of the runner of the *Francis turbine* built at Oodo power plant in 1932, of which rated parameters are: output 3500 kW, head 67 m, flow rate 6.8 m^3/s, speed 450 rpm, and specific speed 154 m-kW. The turbine has suffered cavitation damage and its runner was replaced at intervals of 4–5 years. This runner was installed in 1963, and had been repaired by means of padding every year since 1963. These pictures were taken before it was taken out of service in December 1979.

The state of *cavitation damage* varies with position on the runner as follows:

Slight ruggedness like scales, and deep pin holes scattered about are observed in the region extending from a position at a distance

Fig. 221. Cavitation in a Kaplan turbine (water, rotational speed 1500 rpm, tip speed 196 m/s, runner diameter 250 mm).

of about 6% of the chord length from the leading edge to the maximum thickness. A severely damaged part has grooves stretching radially like the surface of a volcano and is heavily hollowed out as shown in Fig. 219.

There are damage marks like scratches in the flow direction in the vicinity of the maximum thickness, and others like small pipes placed in a row near the trailing edge (Fig. 220).

These different aspects of the damaged surfaces suggest differences in the damaging mechanisms.

Fig. 222. Cavitation on an actual ship propeller (sea water, rotational speed 155 rpm, propeller diameter 3.6 m).

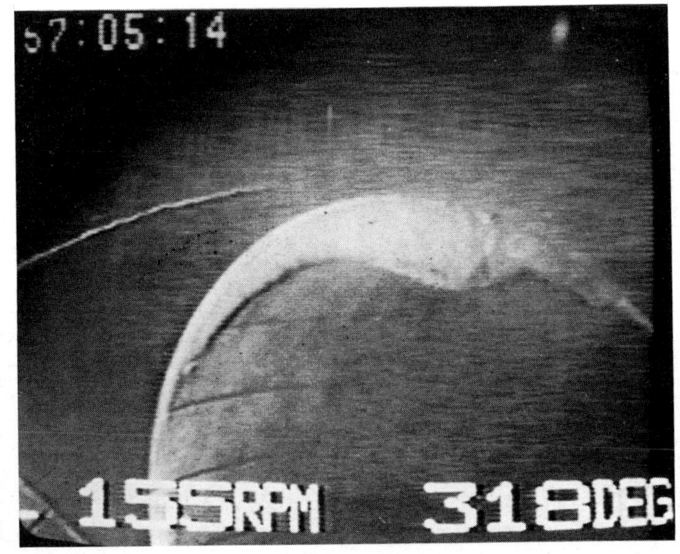

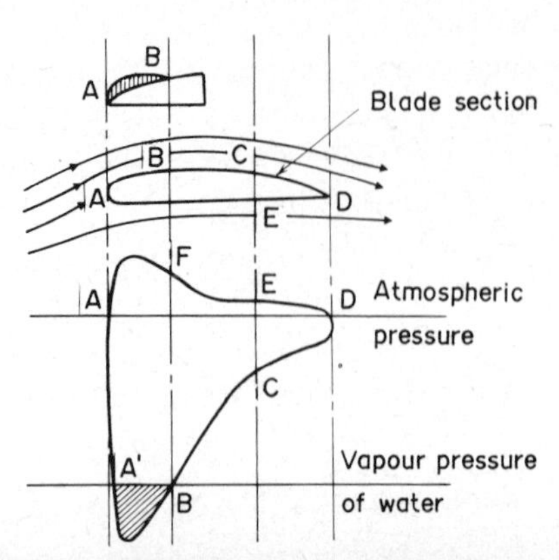

Fig. 223. Cavitation on a ship model propeller
(water, velocity 5.52 m/s, rotational speed 30 rps, propeller diameter 230 mm).

Fig. 224. Cavitation on a ship model propeller
(water, velocity 5.46 m/s, rotational speed 28 rps, propeller diameter 250 mm).

Cavitation in a Kaplan turbine

When cavitation develops in a turbine, sometimes the *run-away speed* becomes higher than when cavitation does not occur. The degree of the speed-up depends on the runner design: therefore, a cavitation test is required for each runner. Figure 221 shows cavitation occurring in *a Kaplan turbine* at run-away speed.

Cavitation on a ship propeller

When one of the blades of *a ship propeller* rotates to the upper part of the propeller disk, cavitation tends to occur because the water head is smaller as the water surface is approached, also a strong wake flow exists there. Figure 222 was obtained using an underwater television camera mounted on the stern of a 105 m-long ship. In the picture, the blade rotates counterclockwise. Cavities look white near the tip and leading edge. A white streak stretching from the left side is a tip vortex cavitation leaving the previous blade.

Cavitation on a ship model propeller

A typical pressure distribution of a rotating propeller blade is shown on the inset. If the pressure between A′–B becomes lower than the vapour pressure of water, cavitation will occur in this region. Figures 223 and 224

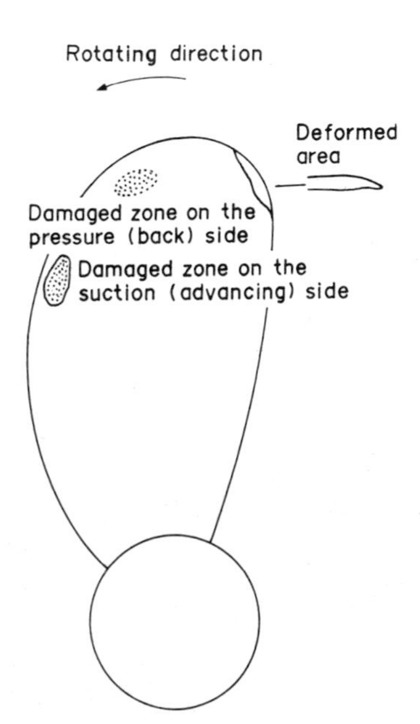

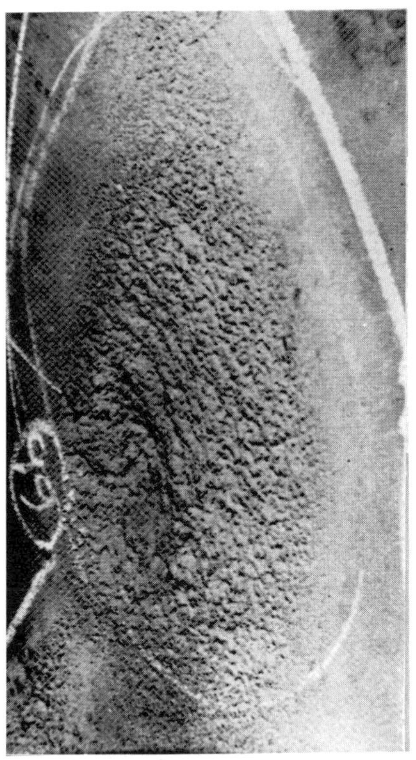

Fig. 225. Cavitation erosion damage to the blade surface of a propeller.

Fig. 226. Deformation of the trailing edge of a propeller blade.

show cavitation on a model propeller photographed with an electronic flash.

When unsteady cavitation occurs on propeller blades rotating in non-uniform wake flow, the cavity volume on each blade changes during one revolution. This is one of the chief causes of hull vibration.

Cavitation damage on a ship propeller

For a merchant ship, cavitation damage appears as deformation and material loss from the backs of the blades, and is usually attributed to the shock of the collapse of cavities as they move into regions of higher pressure toward the trailing edge (Figs. 225 and 226).

Behaviour of a Non-Newtonian Fluid

Weissenberg effect

Polymer solutions and melt, such as polyacrylamide aqueous solution and polyethylene melt, develop normal stresses even in a simple shear flow. This phenomenon is called the *normal stress effect* or the *Weissenberg effect* and it distinguishes this kind of fluid from a Newtonian fluid. Such fluid is termed a *viscoelastic fluid*. Figures 227 and 228 qualitatively show the Weissenberg effect demonstrated by a simple equipment consisting of a beaker with a rotating rod inserted in it. Figure 227 is for the case of a Newtonian fluid (Glycerine aqueous solution). The liquid surface near the rotating rod is pushed outward by centrifugal force. Figure 228 is for the case of the viscoelastic fluid. The liquid climbs up the rod by virtue of the normal stress generated in the fluid.

Secondary flow in a viscoelastic fluid

A Newtonian fluid flows radially from a reservoir into a capillary as shown in Fig. 229. On the other hand a fluid which exhibits the Weissenberg effect develops a large *secondary flow* near the capillary entrance in the reservoir. The secondary flow is stable at a low flow rate as shown in Fig. 230 and increases in size with increasing flow rate. Then, when the flow rate exceeds a certain critical value, the secondary flow becomes unstable as shown in Fig. 231. This phenomenon is also observed in the flow of a polyethylene melt and is a very important consideration in the plastics processing industry. Further this phenomenon causes instability at the capillary exit, leading to the Barus effect and melt fracture of extrudate.

Barus effect (Merrington effect)

When water is drawn from a tube as shown in Fig. 232, the jet contracts at the outlet and the diameter becomes smaller than the tube diameter. But as shown in Fig. 233 the jet diameter becomes larger than the tube diameter for the polymer solution and molten polymer having both viscous and elastic properties. This phenomenon is termed the Barus effect. Experimental studies make it clear that the Barus effect for a short tube is larger than that for longer tube with the same pressure gradient in both tubes. The following have been suggested as the causes of this phenomenon: (i) the elastic strain imposed on entering the capillary is released at the tube exit (the elastic entrance flow effect). (ii) The normal stress generated inside the capillary by the shear flow acts as the recovery force at the tube exit (the normal stress effect). In this connection Metzner et al. have proposed a method for the measurement of stress by using the Barus effect.

Flow on the surface of a rotating disk

Non-Newtonian fluid shows a pecularity of behaviour in shear flow which is not found in a Newtonian fluid such as water or air. The drag in turbulent flow of a polymer solution is markedly reduced under a certain condition. The flow near the surface of rotating disk submerged in a polymer solution is much influenced by centrifugal force, and the angle between the limiting streamline and the circumference is larger than that for the case of Newtonian fluid. The patterns of lines of oil films are different in the two cases due to the difference in the structure of

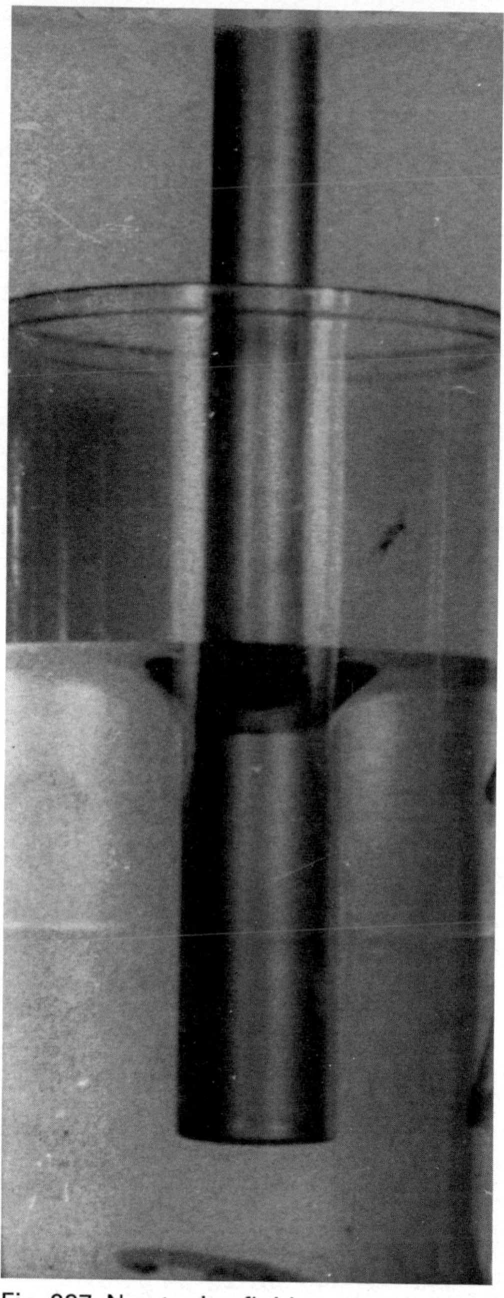

Fig. 227. Newtonian fluid.

Fig. 228. Viscoelastic fluid
(227) glycerine aqueous solution: (228) polyethylene oxide 3% aqueous solution, rotational speed: 700 rpm, rod diameter: 12 mm.

turbulence. Figures 234 and 235 show the flow near the surfaces of rotating disks for water and the polymer solution respectively, made visible by the oil film method. The flow is turbulent except in the vicinity of the centre of the disk. This drag reduction in turbulent flow is called the *Toms effect*.

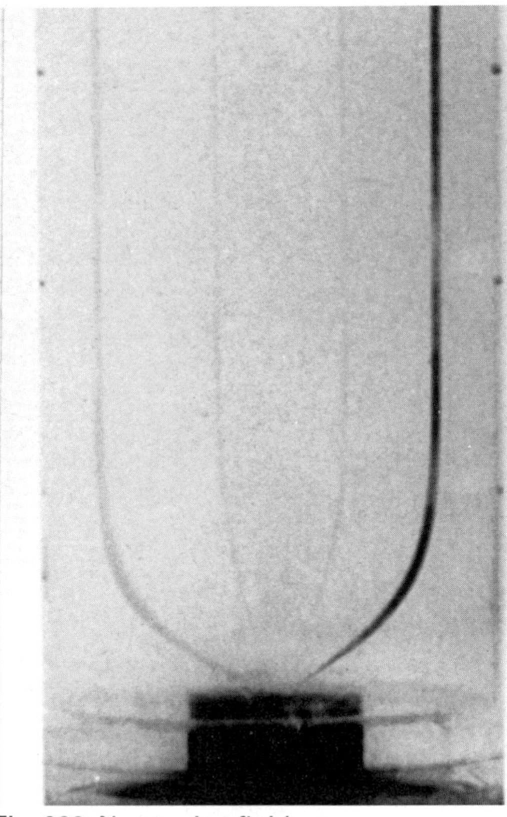

Fig. 229. Newtonian fluid.

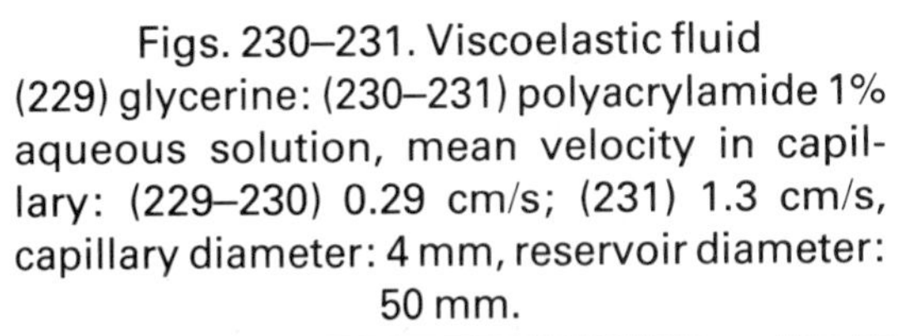

Figs. 230–231. Viscoelastic fluid (229) glycerine: (230–231) polyacrylamide 1% aqueous solution, mean velocity in capillary: (229–230) 0.29 cm/s; (231) 1.3 cm/s, capillary diameter: 4 mm, reservoir diameter: 50 mm.

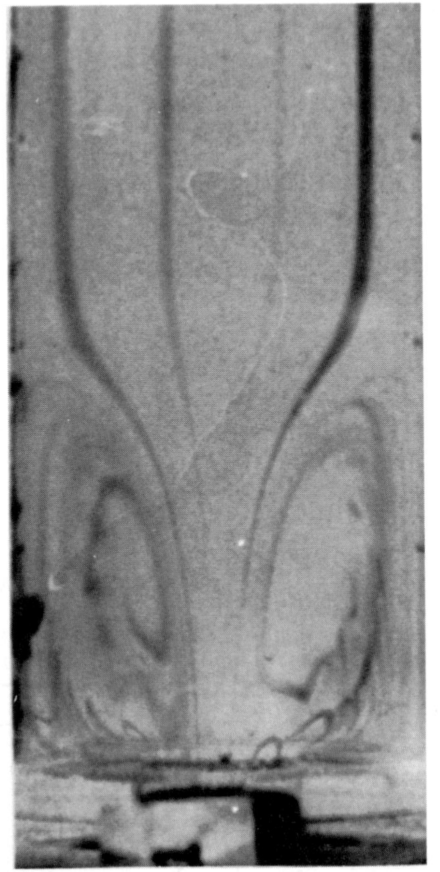

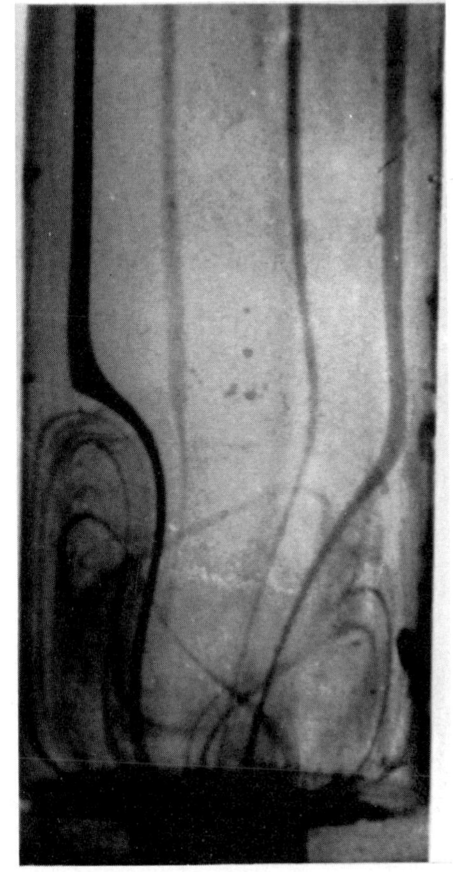

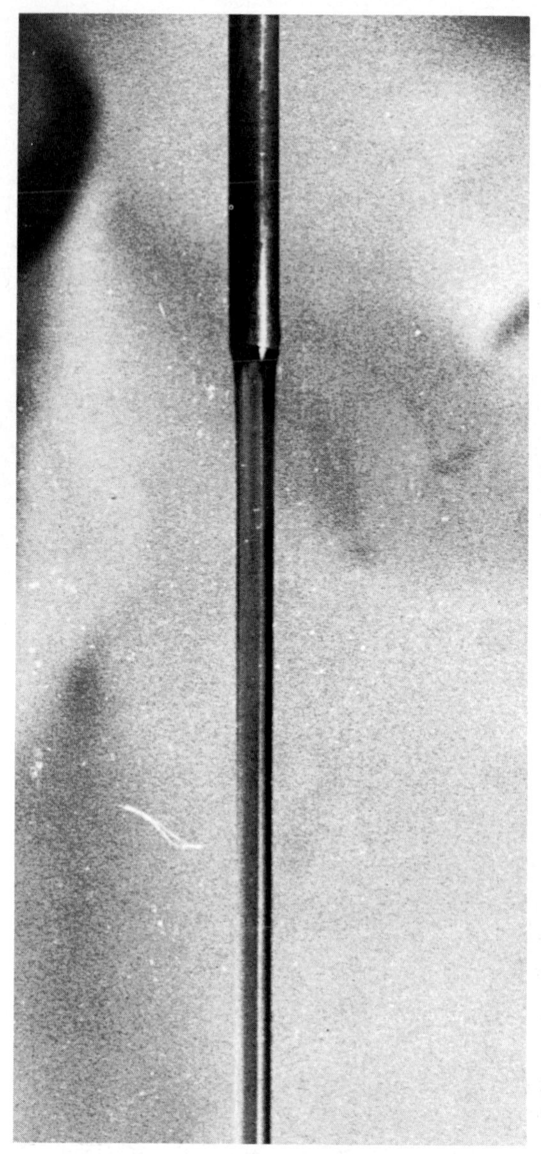

Fig. 232. Water drawn from a tube (water, mean velocity in the tube: 97.4 cm/s, tube diameter: 1.97 mm $Re = 842.3$).

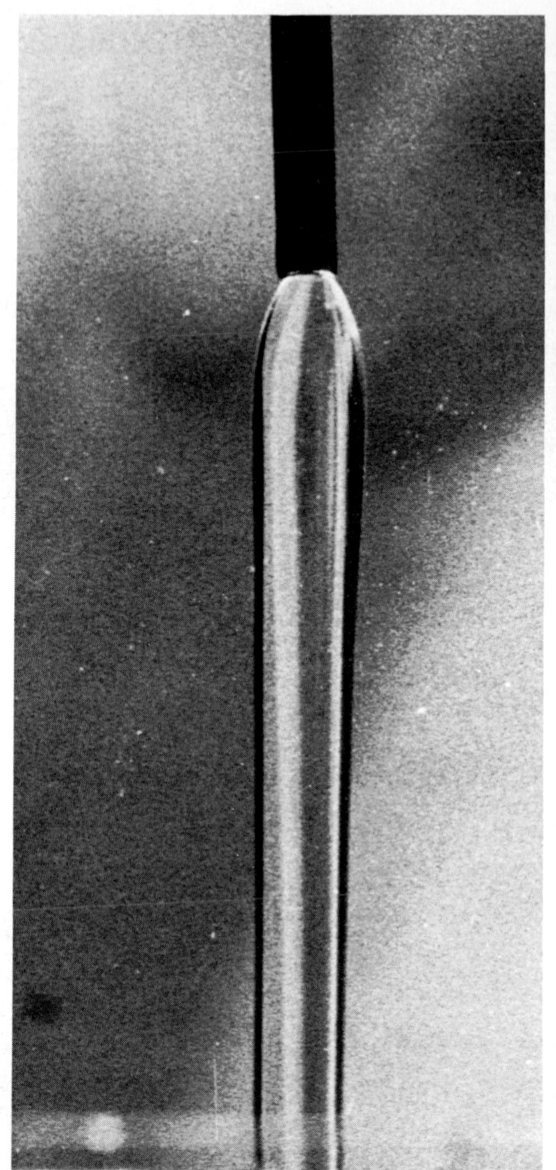

Fig. 233. Viscoelastic fluid drawn from a tube (polymer solution, mean velocity in the tube: 132.5 cm/s, tube diameter: 1.97 mm, $Re = 9$).

Subject Index

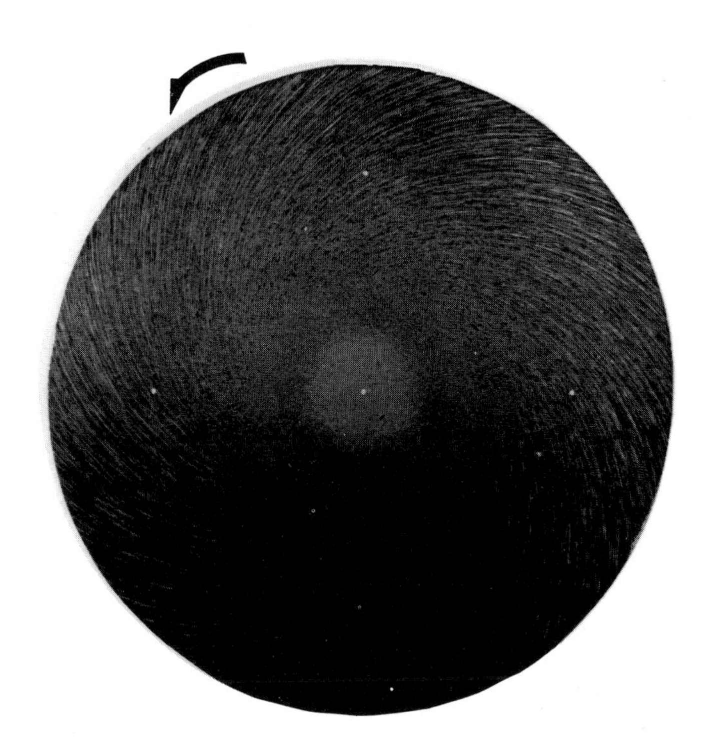

Fig. 234. Flow of water on the surface of a rotating disk.

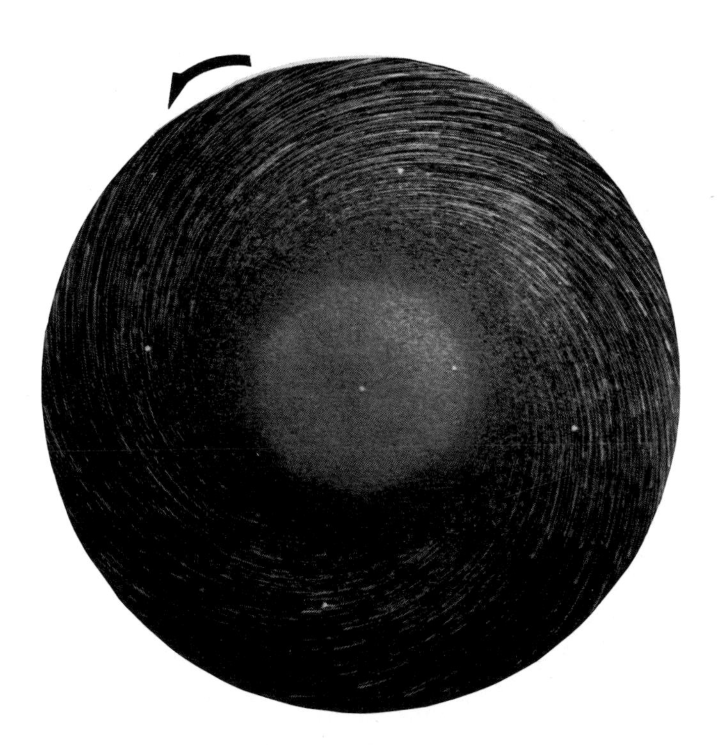

Fig. 235. Flow of polymer solution on the surface of a rotating disk
[(234) water: (235) polyethylene oxide 50 ppm aqueous solution, rotational speed: 800 rpm, disk diameter: 300 mm, $Re = 1.9 \times 10$, oil film method].